FOOD & DRINK –
GOOD MANUFACTURING PRACTICE

A GUIDE TO ITS RESPONSIBLE
MANAGEMENT

Institute of Food Science and Technology (UK)
5 Cambridge Court
210 Shepherd's Bush Road
London W6 7NJ

A John Wiley & Sons, Ltd., Publication

Institute of
Food Science & Technology

This edition first published 2013 © 2013 by John Wiley & Sons, Ltd.
First through fifth edition © The Institute of Food Science & Technology Trust Fund 1978, 1989, 1991, 1998, 2006

Wiley-Blackwell is an imprint of John Wiley & Sons, formed by the merger of Wiley's global Scientific, Technical and Medical business with Blackwell Publishing.

Registered office: John Wiley & Sons, Ltd, The Atrium, Southern Gate, Chichester, West Sussex, PO19 8SQ, UK

Editorial offices: 9600 Garsington Road, Oxford, OX4 2DQ, UK
The Atrium, Southern Gate, Chichester, West Sussex, PO19 8SQ, UK
2121 State Avenue, Ames, Iowa 50014-8300, USA

For details of our global editorial offices, for customer services and for information about how to apply for permission to reuse the copyright material in this book please see our website at www.wiley.com/wiley-blackwell.

The right of the author to be identified as the author of this work has been asserted in accordance with the UK Copyright, Designs and Patents Act 1988.

Library of Congress Cataloging-in-Publication Data is available.
9781118318201

A catalogue record for this book is available from the British Library.

Wiley also publishes its books in a variety of electronic formats. Some content that appears in print may not be available in electronic books.

Cover design: www.hisandhersdesign.co.uk

Set in 9/10.5 pt Times New Roman PS by Toppan Best-set Premedia Limited
Printed and bound in Malaysia by Vivar Printing Sdn Bhd

1   2013

This guide is of an advisory nature. It has been compiled by the Institute of Food Science & Technology in consultation with other interested bodies. It has no statutory force, and nothing in it should be construed as absolving anyone from complying with legal requirements. All possible care has been taken in its preparation, and it is provided for general guidance without liability on the part of the Institute in respect of its application and use.

FOOD & DR
GOOD MAI

A GUIDE TO ITS RESPONSIBLE
MANAGEMENT

Foreword

Good manufacturing processes facilitate improvements in the industry, offering increased protection to consumers and businesses alike. Over the last few years, the EC legislation has been consolidated and simplified, and now sets out, more clearly, the duty of food business owners to produce food safely.

Furthermore, there is growing recognition that underpinning food manufacturing with robust quality assurance procedures can be beneficial within a wider business context and simultaneously improve good practice. I am confident that this will lead to even more efficient practice and better food safety across the sector. Greater confidence in the entire manufacturing process will mean we can have more confidence in the end product.

I am pleased that this latest edition focuses on the need for the industry and senior management to take a more active role in the design, implementation, resourcing and promotion of appropriate and secure food safety processes.

Lord Jeff Rooker
Chair
Food Standards Agency

CONTENTS

Acknowledgements

A list of many of the organisations and individuals from whom help, information or comment has been received for this and previous editions is presented as Appendix V. This is inevitably incomplete and cannot include acknowledgement of numerous verbal comments received. However, I welcome the opportunity to thank all who participated and particularly the members, both past and present, of the GMP Working Groups. Especially, I would thank Professor J.R. Blanchfield, who as Editor and Convener of the GMP Working Group, 4th edition, has made an enormous contribution to the development of the 5th and 6th editions of this Guide.

As with the previous editions, the preparation of this 6th edition has been an enjoyable and enlightening experience.

Louise Manning
Editor, 6th Edition

Preface to the Sixth Edition

The 6th edition has built on previous editions and has focused on the need for every food manufacturing business to have established and implemented a food safety management system (FSMS) appropriate to the products being manufactured, supported and underpinned by the principles of good manufacturing practice (GMP). The adoption of all reasonable precautions relates not only to the measures that have been established in the design of the FSMS and associated quality management system (QMS), but also that these measures are fully implemented and are effective.

There is a requirement therefore to introduce mechanisms for FSMS and QMS validation, and then re-validation as necessary, and for protocols to be established, identifying the measures for monitoring and verification activities. Without these fully functioning and integrated dynamics being in place, due diligence cannot be actively demonstrated.

Louise Manning

Preface to the Fifth Edition

The 5th edition has been developed to meet the various changes in stakeholder requirements of the UK Food Industry since 1998. These include the further development of the European Union (EU) legislation, the establishment of the Food Standards Agency and the Department of the Environment and Rural Affairs [DEFRA to replace the Ministry of Agriculture, Fisheries and Food (MAFF)]. There has also been the development of third-party quality management standards in the food supply chain such as the British Retail Consortium (BRC) Global Standard—Food and establishment of the Global Food Safety Initiative (GFSI) and the introduction of BS EN ISO 22000:2005 Food safety management systems. The 5th edition addresses the need to meet these changes including the requirement for key prerequisite programmes in food manufacture with new or updated chapters on manufacturing activities, cleaning and sanitisation, personnel and training, infestation control, calibration and foreign body controls. Increasing globalisation of food supply chains has required UK businesses to focus on not only product safety, legality and quality but also the need to drive continuous improvement. Sustainable food businesses need to address these issues to produce safe, wholesome food of a consistent quality that meets customer expectations.

L. Manning

Preface to the Fourth Edition

The 3rd edition has served its purpose well and has continued to receive international acclaim. Since its publication, however, there have been important developments, in the intensity of public interest and concern over food safety, and in legislation, not least the adoption of a range of European food hygiene directives and their implementation in the UK in the Food Safety (General Food Hygiene) Regulations 1995 and several parallel Hygiene Regulations affecting dairy products, fresh meat, meat products, poultry and game. Of major significance has been the incorporation of hazard analysis critical control point (HACCP) principles into obligatory European and UK legal requirements. The opportunity has been taken to update the Guide in these and other respects. It is inevitable that legislative references in this Guide concentrate on Europe and the UK; but the principles outlined are of general application and may be interpreted and applied in any country in terms of its own legislation. Increasing interest in novel foods and processes and increased recognition of food allergens as an important food safety issue, have prompted the inclusion of two new chapters on those topics.

J.R. Blanchfield

Preface to the Third Edition

A gratifying level of demand rapidly exhausted supplies of the 2nd edition, necessitating early production of a 3rd edition of the Guide. The opportunity has been taken to review the Guide where appropriate in light of recent European legislative developments such as the EC Official Control of Foodstuffs Directive, and the UK Food Safety Act 1990 and the various regulations and codes of practice made thereunder; to review the texts of existing chapters and to include two new chapters, respectively, on 'Design of Products & Processes' and 'Irradiated Foods'; and to incorporate a number of constructive detailed suggestions received from readers, including some from outside the United Kingdom. Once again, our thanks to our colleagues on the GMP Panel and to all those who have contributed help on this and earlier editions.

K.G. Anderson
J.R. Blanchfield

Preface to the Second Edition

The Institute, and especially those involved in the publication of the first edition of this Guide, are delighted that demand has necessitated a second edition so quickly after its launch in June 1987. We believe that this is indicative of a real need, which the IFST has been able to meet, and we thank again all those who have supported this initiative.

The original GMP Working Group has been replaced by a GMP Panel of the Technical and Legislative Committee, and this Panel has effected a number of corrections and amendments to the first edition, the latter largely of a clarifying nature.

K.G. Anderson (Panel Convenor)
J.R. Blanchfield

Decision Makers' Summary

This summary is especially addressed to the decision makers within food and drink company chairmen, presidents, chief executives, directors and general managers, who are not normally directly involved in detailed design and implementation of good manufacturing practice (GMP) systems, but whose responsibility it is to establish GMP policies and strategies for their companies, and to provide the necessary authority, facilities and resources to the functional managers and staff to implement the requirements effectively.

In this Guide, GMP is considered as that part of a food and drink control operation, which is aimed at ensuring that products are safe to the consumer and are consistently manufactured to a quality appropriate to their intended use. It is thus concerned with both manufacturing and quality management procedures.

The ever-increasing interest among consumers, retailers and enforcement authorities in the conditions and practices in food manufacture and distribution, increases the need for the food manufacturer to operate with clearly defined policies. The ability to demonstrate that the principles and measures identified in this Guide had been fully and effectively implemented, could, in the event of a consumer complaint or a formal prosecution, assist the manufacturer in demonstrating that all reasonable steps had been taken to prevent the cause of the complaint from occurring, or indeed avoid an offence being committed. Enlightened self-interest alone should persuade food manufacturers to follow these guidelines.

The manufacturer of a food product must comply with the relevant legal requirements of the country for which the food is intended, for example, those of composition, of safety, of hygiene and of labelling. While fulfilling these, however, (s)he has a concept of the market at which (s)he is aiming and its requirements (e.g. in the case of a food or drink product, its appearance, flavour, texture, presence or absence or amount of particular nutritional components, inbuilt convenience, shelf life, presentation and price). These factors determine the formulation, processing and packaging of the product.

The product quality as defined by that concept is expressed as a product specification. Conversely, the retailer may approach a manufacturer with a new product concept and request that a manufacturer design a product or process to meet the specific criteria. Of course, the manufacturer's assessment of what the market wants may be correct or incorrect. While the concept effectively meets all of the law's requirements, it may, or may not, effectively meet purchasers' expectations; but unless and until the manufacturer or retailer changes it, it remains the standard with which the product should conform, and GMP is designed to achieve this.

Uniform conformance with product specification is difficult with food and drink products. The main raw materials for food and drink manufacture derive from nature, and are subject to natural variations. In primary production, wide variations may occur among cultivars, and also because of seasonal, weather and cultivation differences. In animals, apart from variations among and even within individuals, differences between breeds and rearing systems lead to variations.

Therefore the additional task of the food or drink manufacturer, aided by the knowledge and skills of food science and technology, is to make a reasonably uniform product from variable raw materials by an appropriate combination of raw material selection, raw material pretreatment, formulation adjustment and processing variation.

GMP has two complementary and interacting components: the manufacturing operations and the quality management system [which, for the purposes of this Guide, the Institute of Food Science & Technology (IFST) has designated 'food control'] (see Figure 1). Both these components must be well designed and effectively implemented. The same complementary nature and interaction must apply to the respective management of these two functions, with the authority and responsibilities of each clearly defined, agreed and mutually recognised. This is not to disregard the importance of other key functions essential to the effective functioning of a company, or indeed of those functions contributing direct services or advice to the manufacturing operation (e.g. purchasing, cost accounting, work study, production planning and engineering maintenance).

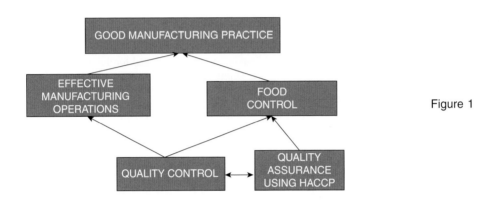

Figure 1

What constitutes 'well designed' in these two contexts mentioned above is not just a matter of common sense, or something that would be self-evident to non-technical business people. As well as management skills, it also involves extensive and up to date knowledge of current and emerging food safety hazards and food science and technology as relating to the ingredients, processes, packaging and products concerned.

Effective Manufacturing Operations

GMP requires that every aspect of manufacture is fully defined in advance; and that all the resources and facilities are specified—namely:

- specific measures undertaken at critical control points (CCPs) based on food safety hazard analysis;
- adequate design of premises and suitable manufacturing and storage space;
- suitable process flow with process design to streamline the process and minimise the potential for cross-contamination;
- correct and adequately maintained equipment;
- appropriately trained people;
- correct raw materials, processing aids and packaging materials;
- appropriate storage and transport facilities;
- documented operational procedures and cleaning schedules;
- appropriate management and supervision; and
- adequate technical, administrative and maintenance services

are provided, in the right quantities, at the right times and places, and are utilised as intended. In order to ensure that operations do proceed according to plan, it is also necessary to:

- provide operators with documented procedures in clear unambiguous instructional language (with due regard to reading, numeracy and language problems);
- train and motivate the operators to carry out the procedures correctly;
- undertake formal review to ensure that training has been effective;
- avoid, if possible, incentive bonus schemes, but, if unavoidable, to build into any incentive bonus scheme adequate safeguards against unauthorised 'short cuts';
- provide a food control programme working along the lines indicated below;
- ensure that genuine records are completed during production and that they demonstrate that specified procedures were in fact complied with, and to enable the history of manufacture and distribution of a batch subsequently to be traced should a problem arise or a product withdrawal or recall be necessary;
- establish a well-planned and effective system to carry out a product withdrawal or recall, should that prove necessary; and
- establish a tried and proved business continuity and crisis management procedure in case of need.

Effective Food Control

The other and complementary major component of GMP is effective food control. Effectiveness requires:

- well-qualified and appropriately experienced food control management participating in the development and validation of process controls and specifications;
- competent staff and adequate facilities to do all the relevant inspection, sampling and testing of materials, and monitoring of process conditions and relevant aspects of the production environment (including all aspects of hygiene) and management of potential food safety hazards;
- verification activities to be developed and implemented by appropriately experienced personnel in order to demonstrate that the food products and the process are under the appropriate level of control; and
- rapid feedback of information (accompanied where necessary by advice) to manufacturing personnel, thereby enabling prompt adjustment or corrective action to be taken, and enabling processed material to be approved as fit for either further processing or sale, or to be segregated for decision as to appropriate disposition, for example, reject, regrade or reprocessing.

Responsible Management

Of course, the requirements of effective manufacturing operations and of effective food control mentioned above are merely headings; and within each there are very many aspects that are considered more fully within the body of this Guide. The Institute hopes that the Guide will prove of help to the management of food and drink companies, to those concerned with private and public verification activities, food law enforcement and consumer protection, to the students who will be the food technologists, engineers and production managers of tomorrow and to those responsible for training them.

The full title of the Guide is 'Food & Drink – Good Manufacturing Practice: A Guide to its Responsible Management'. The reference to responsible management is deliberate. GMP can only stem from policy firmly and uncompromisingly stated and continuously pursued by a company board and general management, which, moreover, provides adequate physical, financial and human resources for the purpose.

Food & Drink –

Good Manufacturing Practice

A Guide to its Responsible Management

PART I – GENERAL GUIDANCE

1   INTRODUCTION

1.1   The purpose of this Guide is to outline the responsibilities of managers in relation to the efficient manufacture and control of food and drink products, thereby ensuring that such products are safe, wholesome and of the nature and quality intended. While it addresses manufacture of food and drink for use in the catering and vending industries, it does not deal with catering and retail activities per se. The Guide is therefore particularly concerned with advice to management on:

- matters affecting product safety, including health and hygiene of personnel relating thereto;
- product manufacture in terms of product and process control and handling under hygienic conditions in conformity with product, packaging and labelling specifications; and
- associated matters such as training of personnel, documentation and record keeping, supplier approval, suitability of premises and equipment and site standards, waste avoidance, recovery and reworking of materials, laboratory management, traceability, verification activities, and preventive and corrective action and the management of customer complaints and product recall.

1.2   It is emphasised that the Guide is concerned with advice on principles, and it is recognised that methods other than those described, but which achieve the same ends, may be equally acceptable. Personnel and premises hygiene, because of its importance, is treated as a continuous theme and a subject for consideration throughout the document.

The Guide is in three parts:

Part I: deals with matters of general application;
Part II: deals with guidance on specific manufacturing and/or food categories; and
Part III: covers mechanisms for review of the Guide.

Food & Drink – Good Manufacturing Practice: A Guide to its Responsible Management, Sixth Edition.
The Institute of Food Science & Technology Trust Fund.
© 2013 John Wiley & Sons, Ltd. Published 2013 by John Wiley & Sons, Ltd.

1.3   The Guide does not deal directly with such matters as operative safety and welfare, ethical matters, animal welfare or environmental issues including water and energy conservation. It refers to resource management and waste control, engineering, maintenance and transport and distribution only in respect of those aspects that have a bearing on product safety and integrity. In general it does not deal with matters unrelated to scientific, technological and organisational aspects affecting product quality and safety.

1.4   **EU Regulation (EC) 853/2004**, laying down specific hygiene rules for the hygiene of foodstuffs, requires Approval of any establishment producing foods of animal origin. Any change in production may affect the status of such Approval. This includes (although not exclusively so) changes to product range and to product descriptions, changes in the production process, changes in equipment and in establishment design and layout. In such cases re-approval may be required and FBO's should contact their Local Food Authority.

The Leading case **Allan Rich Seafoods v. Lincoln Magistrates' Court [2009] EWHC 3391** confirmed that changes in the Food Business Operator (FBO) also require re-approval of such an establishment.

Any change in production may affect the status of Approval under Regulation (EC) 853/2004. This includes (although not exclusively so) changes to product range and to product descriptions, changes in the production process, changes in equipment and in establishment design and layout. In such cases re-approval may be required and FBO's should contact their Local Food Authority.

1.5   The Guide will also make reference to international standards such as those developed by the Codex Alimentarius Commission. It is then the responsibility of the reader to refer to current legislation itself or review with the support of a competent adviser, and not to rely on an interpretation or an abridged version of the requirements as given in this document.

In addition to mandatory legal requirements, the Guide is concerned with advisable practices some of which may already be contained in published guidelines or codes of practice. The Guide outlines general principles and gives references that the reader is advised to consult in full.

1.6   The initial adoption of the **EC Official Control of Foodstuffs Directive** and the advent of the **UK Food Safety Act 1990** as well as existing provisions of the **UK Trade Descriptions Act** and the **UK Weights and Measures Act** gave increasing emphasis to the need for a manufacturer to be able to prove that (s)/he did everything necessary to comply with the law. Thus under the **Food Safety Act**

1990, and other subsequent legislation such as the **General Food Regulations 2004** and the **Food Hygiene Regulations 2006**, a manufacturer, retailer or importer charged with an offence may enter the legal defence that (s)/he **'took all reasonable precautions and exercised all due diligence to avoid the commission of the offence by himself or by a person under his control'**. In this context, it can be considered that 'precautions' are the measures taken and 'diligence' is the activities undertaken to ensure their effective application. The wording puts the onus of proof on the defendant, and both must be proved and the use of the word 'all' implies that 'some' or 'most' will not be enough. What constitutes 'all reasonable precautions and all due diligence' in a particular instance must relate to the nature of the offence and to other related circumstances. Nevertheless in the case of a safety or a 'nature, substance or quality' offence, a manufacturer who can **prove** that (s)/he has diligently installed and appropriately applied all the relevant measures in the Institute of Food Science & Technology (IFST) Guide to Good Manufacturing Practice (GMP) will stand a very good chance of a successful defence. It must also be pointed out that a manufacturer who does not employ appropriate technically competent personnel to specify the product formulation, factory processes and procedures, to design and control the continuous monitoring of their correct operation and undertake such validation and verification activities cannot be said to have exercised either adequate precautions or adequate diligence and is unlikely to have a successful defence.

1.7 Responsibility for enforcement within the EU varies from country to country. In the UK it is shared between central and local government bodies. While the making of legislation in the UK is the function of central government, the enforcement of food law is primarily (but not solely) the responsibility of more than 400 local authorities (LAs) in the UK, and more specifically environmental health officers (EHOs) and trading standards officers (TSOs).

The responsibilities of all the enforcement authorities in the UK are set out in the Single Integrated National Control Plan for the UK; this demonstrates the complexity of structures of enforcement across the UK, and the paragraphs below are a broad overview for those in food manufacturing and arrangements may be different within your geographical area or industry sector.

The Food Standards Agency (FSA) interaction with enforcement officers is set out in the Framework Agreement on Local Authority Enforcement. This document gives a structure to the Agency's supervision of LA enforcement work.

The Food Law Code of Practice (FLCP) sets out the way LAs should apply food law, and how they should work with food businesses. Practical guidance is also provided as a further help to enforcement officers.

The EHOs and TSOs have to be authorised by their LAs to enforce the food legislation. Once they achieve certain qualifications, detailed under the FLCP, they are authorised to carry out certain tasks and are provided with powers (under the Food Safety Act 1990) to, for example, enter premises, take samples, gather evidence, issue notices and, under certain circumstances, close premises.

Depending on the structure of the Local Government in the area in England and Wales, food visits may be from TSOs to examine labelling, compositional standards and food contaminants, and EHOs to check on food hygiene; however in Scotland, Northern Ireland and some Welsh and English authorities, EHOs are responsible for all the food legislation with TSOs responsible for weights and measures checks. Visits to manufacturing sites are to ensure compliance with legislation; the frequency of interventions (visits) is assessed according to the risk assessment system detailed in the FLCP, and most manufacturing units can expect a visit at least once a year.

The actual policy and resources allocated to the inspection premises and sampling of product will depend on the LA and therefore there are variations in delivery. However businesses should be able to benefit from a positive relationship with enforcement authorities receiving detailed written feedback following inspections and receive results of sampling exercises. Companies who develop a 'Home Authority' or 'Primary Authority' agreement with their LAs can expect a more detailed and possibly, a more supportive relationship with the benefits of the Better Regulation agenda.

1.8   Absolute terms, such as 'ensure that', 'avoid', 'prevent', 'absence of . . . ' and so on, have been used in various parts of the Guide. To dispense with them would detract from the intentions of the Guide or would necessitate lengthy explanations on each occasion. Accordingly, readers should note that such terms are to be interpreted in a rational and practical way, for example, 'ensure that . . . ' should be read as meaning 'ensure, so far as is reasonably practicable, that . . . '. Words such as 'should' are used for non-mandatory advice, and the imperative, for example, 'must' or 'shall' is reserved for appropriate mandatory requirements.

1.9   Definitions of some of the terms used in this Guide are given in Appendix I. It is appreciated that other definitions may be equally valid or preferred, and the appendix definitions are simply intended to clarify the meanings attributed to a word or phrase when used in the compilation of the Guide.

1.10   The Guide is an advisory document with a list of supplementary references intended to assist all grades of management. It may be particularly useful to students studying food manufacture, to new entrants to management and to general managers in smaller companies who may be responsible for a range of management functions each of which may be the sole concern of one or more specialist senior managers in a larger company as well as regulatory officers.

1.11   GMP is not a static concept, but an evolutionary, dynamic mechanism by which overall improvements can be made and maintained.

1.12   Abbreviations, for example, GMP, have been used in the text but have been reconfirmed at the start of each chapter in case the chapter is read in isolation and therefore to minimise the number of times that the reader has to refer to the abbreviations list (Appendix II).

Principle

There should be a comprehensive quality management system (QMS), so designed, documented, implemented, reviewed and continuously improved, and so furnished with personnel, equipment and resources, as to ensure that specifications set to achieve the intended product quality standards are consistently met. The attainment of this quality objective requires the involvement and commitment of all concerned at all stages of manufacture.

Explanatory Note   2.1   A manufacturer has to comply with the legal requirements relevant to the product manufactured. While embracing these, (s)he will have determined the market requirement that (s)he aims to meet, and therefore the product quality standard. The established product specification embodies both legal requirements (e.g. those of composition, safety, hygiene and labelling) and market requirements (such as product nature, appearance, flavour, texture, presence or absence and quantity of particular nutritional components, nature of pack, pack size, degree of inbuilt convenience, shelf life, presentation and price). While some commercial and marketing considerations affecting the market requirement specification are outside the scope of this Guide, those relating to the principles of design and development of products and processes to comply with that specification are dealt with in Chapter 8. The product and process design, when completed and validated, then becomes a part of the full product specification. Once established it remains permanent until formally changed. All references in this Guide to compliance with product specifications imply compliance with all of the foregoing requirements embodied in the specification.

2.2   In order to achieve the objectives of good manufacturing practice (GMP), it is necessary to have in place:
1. *Quality Assurance:* that is, to design and plan, as relevant, raw material specifications, ingredients formulation, adequate resources such as processing equipment and environment, processing methods and conditions, intermediates specifications, appropriate packaging and labelling specifications, specification for quantity per pack, specifications for management and control procedures, a specified distribution system and cycle, and appropriate storage, handling and preparation instructions, which, taken all together, are capable of resulting in products complying with the product specification;

2. *Effective Manufacturing Operations:* that is, to validate and manage the operational production/distribution practices so as to ensure that the capability is translated into reality, so that, the

Food & Drink – Good Manufacturing Practice: A Guide to its Responsible Management, Sixth Edition.
The Institute of Food Science & Technology Trust Fund.
© 2013 John Wiley & Sons, Ltd. Published 2013 by John Wiley & Sons, Ltd.

process adheres to its specified design parameters and that the resulting products actually do consistently comply with the product specification. This is relevant for quality, legislative and food safety criteria;

3. *Quality Control:* that is, to have in place an effective monitoring system that checks compliance with specified requirements and defines suitable corrective action in the event of 'out-of-control' occurrences. Furthermore an effective verification system is required that confirms that the quality plan and associated quality activities will consistently deliver products of the specified quality at the point of production and throughout the product's shelf life.

Good Manufacturing 2.3
Practice (GMP)

Thus, GMP may be viewed as having two complementary components, namely effective manufacturing operations and effective food control (see Figure 1).

Food Control 2.4

The Institute of Food Science & Technology (IFST) uses the term 'food control' to describe a comprehensive QMS and an integrated food safety management system (FSMS) based on the principles of hazard analysis critical control point (HACCP). It is vital that the FSMS and associated prerequisite programmes (PRPs) interlink with food safety validation and verification and quality assurance and quality control within a QMS, which is appropriate to the products and processes involved and the inherent level of food safety risk.

Quality Management 2.5
Systems

Many manufacturers will have developed their own QMS, but increasingly are attaining or seeking to attain certification to a private QMS standard. EN ISO 9000:2000 is an international QMS standard, and it describes the requirements of a QMS to assure conformance of product and production to specified requirements. Furthermore, QMS standards have been developed specifically for food manufacturing especially with the increasing globalisation of food production. These include BS EN ISO 22000:2005, the British Retail Consortium (BRC) Global Standard for Food Safety and the Global Food Safety Initiative (GFSI) Standard developed to benchmark international private standards.

Effective 2.6
Manufacturing
Operations

An effective manufacturing operation is one where, as appropriate:

(a) the manufacturing process, equipment, activities, precautions and so on are fully specified in advance, and systematically reviewed in light of experience;
(b) the necessary facilities and resources are provided, including:
   (i)   appropriately qualified personnel,
   (ii)   adequate premises and space,
   (iii) suitable equipment and services,
   (iv)   specified materials, including packaging,
   (v)   specified policies and procedures, including cleaning procedures, and
   (vi)   suitable storage and transport;

(c) the relevant written procedures are provided in instructional form and using clear and unambiguous language, and are specifically applicable to the facilities provided;

(d) operators are trained and motivated to carry out the procedures correctly, and refresher training is undertaken at appropriate intervals;

(e) records are made (whether manually or by recording instruments or both) during all stages of manufacture, which demonstrate that all the processing steps required by the defined procedures were in fact carried out, and that the quantity and quality of product produced were those expected;

(f) records are made and retained in legible and accessible form, which enables the history of the manufacture and distribution of a batch to be traced;

(g) a system is available to withdraw or recall from sale or supply any batch of product, if that should become necessary; and

(h) a review system is in place to consider actual operational performance against proposed performance and drive the implementation of appropriate preventive and corrective action where appropriate.

Effective Quality Control 2.7 Quality control is the function concerned with determining the compliance of the finished products with specifications and with activities ancillary thereto. It includes the undertaking of inspections and tests to determine the degree of compliance with specifications, the examination of process control data and the provision of rapid information and advice leading to corrective action when necessary. It is therefore a 'lag' activity designed to detect product and process failure rather than in the case of quality assurance activities to prevent product and process failure. The term is also used to designate the department responsible for this function. (N.B. What is described below does not preclude automatic process adjustment by negative feedback from automatic process monitors/recorders, or production operators receiving such information on-screen and then taking appropriate action, provided that they are suitably trained, and that such procedures are written into the quality control system and that any actions undertaken by personnel are recorded.) In describing the role of the quality control manager below, it is recognised that alternative job titles may be used by organisations, but it is important for all food manufacturing organisations to distinguish clearly the management roles of quality assurance (failure prevention) and quality control (failure detection) especially where they are effectively managed by the same person.

Effective quality control requires that, where appropriate:

2.8 (a) the quality control manager participates, with others as necessary, in the assurance role of development and approval of specifications, liaising with suppliers in agreeing product specifications and service requirements, and the control function of assessing and approving suppliers on the basis of their ability on an ongoing basis to supply reliably in compliance with the specifications;

(b) adequate resources, facilities and staff are available for sampling, inspection, testing and sensory assessment of starting materials (including packaging materials), intermediates and finished products, and for monitoring process and storage conditions and relevant aspects of the production environment (including all aspects of hygiene);

(c) all samples for inspection and testing are representative of the batch being sampled, are collected by personnel under the direction of, and examined with methods approved by, the quality control manager. The results of such examination need to be formally assessed against the specification by the quality control manager or a competent person designated by him/her;

(d) established procedures exist whereby starting materials and intermediates are approved for use, rejected or designated for treatment intended to bring them within specification, according to inspection/test results obtained;

(e) there is rapid feedback of information (accompanied where appropriate, by advice) to manufacturing personnel, enabling prompt adjustment or corrective action to be taken when necessary, and to the purchasing function in respect of raw material lots;

(f) a positive release procedure exists, where appropriate, whereby batches of finished products are temporarily quarantined until formally released for rectification, or into normal stock, or for distribution;

(g) sufficient reference samples of starting materials, or records of the result of their inspection where deterioration could occur, should be retained to permit future examination if necessary;

(h) sufficient reference samples of finished products are retained for shelf-life tests and to permit future examination if necessary;

(i) customer/consumer complaint samples are examined, the causes of defects are investigated where possible and appropriate measures are advised for corrective action to prevent recurrence;

(j) summaries of quality performance data in appropriate form are provided by quality control to operating functions (e.g. general management, production management, purchasing and cost accounting). These summaries may provide input in the determination of quality objectives for the business whereby data are routinely analysed to determine performance against defined targets and potentially identify areas for improvement;

(k) a direct interest is taken in the activities and quality assurance procedures of the suppliers of raw materials and packaging materials, and close contact is maintained with their quality assurance departments;

(l) ongoing contact is maintained with the relevant enforcement authorities and matters raised by them are investigated and responded to; in the UK the Food Standards Agency (FSA) and the 'Home Authority' will provide useful contacts;

(m) due heed is taken of new developments in food legislation, especially on changes in compositional standards and labelling requirements that may necessitate changes to specifications for raw materials or finished products, and on European Union (EU) and UK Government proposals for future food legislation; and

(n) the authority and responsibilities of the production management and the quality control management functions respectively are clearly defined so that there is no misunderstanding (see Chapter 11).

3 HAZARD ANALYSIS CRITICAL CONTROL POINT (HACCP)

Principle

There should be a comprehensive food safety management system (FSMS), so designed, documented, implemented and reviewed, and so furnished with personnel, equipment and resources, as to ensure that critical limits set to achieve the intended food safety standards are not exceeded. The attainment of this food safety objective requires the design, development and implementation of a hazard analysis critical control point (HACCP) system specific to the manufacturing process and the commitment of all concerned at all stages of manufacture.

Hygienic Practice and Prerequisite Programmes (PRPs)

3.1 Good hygienic practice is critical to every aspect of good manufacturing practice (GMP), and throughout the Guide it has been treated as a continuous theme and has deliberately not been made the subject of a separate chapter. The Codex Alimentarius Commission (CAC) recommended international code of practice General Principles of Food Hygiene CAC/RCP 1-1969 (2003; Rev 4) lays down the foundation for ensuring food hygiene, and key aspects are addressed in this Guide. The term PRP is often used to identify the procedures, policies and protocols that need to be in place within a food organisation before a HACCP plan can be designed and implemented. A number of these requirements are detailed in the previously mentioned CAC/RCP. Examples of PRPs include personal hygiene protocols, premises hygiene procedures, calibration, training and pest control programmes. The PAS 220:2008 standard was designed to assist organisations seeking to establish, implement and maintain PRP in order to meet the elements of BS EN ISO 22000:2005. PAS 220:2008 was updated and has now become ISO/TS 22002-1:2009 Prerequisite programmes on food safety—Part 1: Food manufacturing.

The 'hygiene package' of five laws adopted by the European Union (EU) in 2004 aimed to merge, harmonise and simplify the complex hygiene requirements that were hitherto contained within seventeen EU Directives. The aim was to create a simple, transparent hygiene policy applicable to all food and all food operators together with effective instruments to manage food safety and food safety management throughout the supply chain. The new hygiene law has applied since 1 January 2006.

HACCP

3.2 With regard to current legislation in the EU, during the design and implementation of manufacturing operations and control procedures, **HACCP principles must be applied as defined in the EU Regulation (EC) 852/2004 of the European Parliament and of The Council, in which Regulation 1 requires:**

Food & Drink – Good Manufacturing Practice: A Guide to its Responsible Management, Sixth Edition.
The Institute of Food Science & Technology Trust Fund.
© 2013 John Wiley & Sons, Ltd. Published 2013 by John Wiley & Sons, Ltd.

- **general implementation of procedures based on the HACCP principles, together with the application of good hygiene practice, should reinforce food business operators' responsibility;**
- **guides to good practice are a valuable instrument to aid food business operators at all levels of the food chain with compliance with food hygiene rules and with the application of the HACCP principles.**

Regulation 2 (a) to (g) defines those HACCP principles. An EU Regulation has immediate force on the due date in all Member States. Provisions for enforcement and penalties in the UK are contained in the Food Hygiene (England) Regulations 2005 and similar Regulations for Scotland, Wales and Northern Ireland.

3.3   It takes more than common sense or business acumen to be able to comply with these legal requirements. In large- and medium-sized food business establishments, it requires suitable numbers of appropriately qualified and experienced personnel. Even in the smallest food business, it is extremely important that the proprietor or some other responsible person has been trained in the principles of food hygiene and food safety, at least to Level 3 standard. There must be senior management commitment to HACCP, which will be implemented through the FSMS.

Although food safety is the most important factor, to which the application of the above principles is mandatory, the principles are also applicable to preventing or minimising defects in respect of quality attributes.

3.4   The HACCP system and guidelines for its application is published in the Codex Alimentarius Commission Food Hygiene Basic Texts ISBN 92-5-104021-4 and identifies seven principles of HACCP:
1. Conduct a hazard analysis. Prepare a list of steps in the process where significant hazards can occur and describe the preventive measures.

2. Identify the critical control points (CCPs) in the process.

3. Establish critical limits for preventive measures associated with each identified CCP.

4. Establish CCP monitoring requirements. Establish procedures for using the results of monitoring to adjust the process and maintain control.

5. Establish corrective actions to be taken when monitoring indicates that there is a deviation from an established critical limit.

6. Establish effective record-keeping procedures that document the HACCP system.

7. Establish procedures for verification that the HACCP system is working correctly.

3.5   In order to undertake hazard analysis or HACCP, a team should be drawn together. The HACCP team needs to contain personnel who have expertise in areas such as production, engineering, quality control, product technology and procurement. The team members need to have relevant practical experience, knowledge of the products and processes within the study and suitable training on how to undertake a HACCP study and the implementation of HACCP principles. At least one member of the team should have formal HACCP training, but all team members need to be trained on how to utilise HACCP principles in assessing how a food product should be manufactured in order to minimise the potential for a food safety incident occurring. The team is also responsible for ongoing review and management of the HACCP system. In the event that external expertise is sourced to assist with either the development or the maintenance of the HACCP system, it is critical that the management team should not delegate responsibility to the external resource. The management of the HACCP system and the development and implementation of the food safety control system remain the responsibility of the manufacturing organisation. The quality of the external expertise should be formally assessed including the amount of experience in the food industry and the provision of appropriate references from current clients.

3.6   The scope of the HACCP plan(s), that is, the products produced and processes undertaken at the manufacturing site, should be detailed. Information about the food product is usually recorded in a product specification. Product specifications should be reviewed to ensure that they contain the relevant information. This should include:

(a) product composition in terms of ingredients, including the origin of ingredients, nature of the item in the case of fruit or vegetables, whether or not the product contains allergens;
(b) the physical and chemical attributes of the food including those that might limit microbial growth, for example, salt or sugar content, pH or water activity;
(c) packaging type and standards, for example, gas modified atmosphere or vacuum packed;
(d) storage and distribution requirements;
(e) instructions for use;
(f) intended consumer target group, for example, the general population or a specific group that may be more vulnerable to the food safety hazards being assessed; and
(g) shelf life and nutrition information.

The nature of the treatment and processing of the ingredients and final product should also be defined in the product specification, or an alternative document. This is especially so where process activities are specifically designed to reduce the likelihood of a food hazard occurring, or surviving the processing treatment, for example, heat treatment and foreign body detection.

3.7   HACCP is essentially a preventive methodology that needs to be exercised not only within the confines of the in-factory manufacturing process. It should also be applied to the sourcing and intake of the starting materials and packaging materials, and to the post-process packaging, handling and distribution, and indeed, as far as possible, via appropriate storage, preparation and use instructions on the label, as far as the consumer.

3.8   A process flow diagram needs to be developed to identify each step within the manufacturing process. BS EN ISO 22000:2005 describes a flow diagram as a 'schematic and systematic presentation of the sequence of, and interaction of steps' and states that flow diagrams should be prepared for the process(es) and product(s) within the scope of the HACCP or FSMS. Flow diagrams should include as applicable the sequence of steps from intake through each definable stage, to intermediate and finished products, despatch and delivery to the consumer, to stages where reworking, regrading or recycling takes place and where waste is produced.

Verifying the flow diagram involves physically walking the flow diagram in the manufacturing premises. The 'walking' of the process flow diagram is important to identify potential hazards that have not been identified in the initial review stages; to determine the degree of implementation of PRPs and preventive (control) measures in practice; to identify areas of potential cross-contamination; to determine holding periods for product especially as a result of equipment breakdown and if these could be to the detriment of the product; and to determine whether all process steps (both forward and back in the case of rework loops) have been included in the flow diagram. This verification activity will also aid the determination of realistic food safety hazards. Verification of the site layout plan should also be undertaken at the same time. Records of verification should be maintained and the frequency of verification should also be agreed. Re-verification activities ensure that any changes to the process flow diagram or the site layout plan have been adequately recorded, and re-verification activities should be scheduled at designated intervals.

3.9   As well as the development of a schematic flow diagram that outlines the individual process steps in preparing, storing, manufacturing and despatching the product, further factors should be considered. A site layout plan for the site as a whole, and the manufacturing areas specifically, should be developed. This plan should cover internal and external areas. It should identify people, product and process flow especially where there is the potential for delay, rework or recycling. It should also include the availability of and access to utilities such as water, ice and air, especially in the instance where there may be both a potable and non-potable supply of water. This plan should also identify segregation by area, for example, allergen control, low-care/high-care areas or low-risk/high-risk areas, depending on the products being stored and the manufacturing process being undertaken. The location of waste, drainage systems and cleaning chemical storage should also be identified as well as the flow of waste to external storage. This plan should be used when considering the potential for contamination with extrinsic hazards. In this context, extrinsic hazards would be deemed to be hazards arising from people, the manufacturing environment, waste and/or other products being manufactured. Process design should be reviewed in order to ensure there is the minimum potential for cross-contamination of this nature. This review should also consider external access and the security requirements in terms of product risk especially where items are stored in external locations.

3.10   The hazard analysis should consider all realistic potential hazards that could occur at each stage of the manufacturing process and the potential cause. Classic hazard analysis defines three types of food safety hazard: biological (otherwise called microbiological), chemical and physical. This basic classification of food safety hazards needs to be set in the context of emerging hazards and further hazard types being identified, for example, food allergens that have the potential to cause an allergic reaction when handled or consumed.

The CAC Guidelines for the Validation of Food Safety Control Measures (CAC/GL 69—2008) stated that the control of hazards potentially associated with foods usually involves the application of control measures in the food chain, from primary production, through processing, to consumption. The guidelines describe a control measure as any action and activity that can be used to prevent or eliminate a food safety hazard or reduce it to an acceptable level. In this context the terms preventive measure and control measure can be considered to be interchangeable.

For each realistic hazard, analysis is required to take account of the severity of the hazard and the likelihood of it occurring and whether elimination or reduction to an acceptable level is critical to ensure food safety. Account should be taken of subsequent stages in the production process and their potential impact on eliminating or reducing the hazard to an acceptable level and hence the impact any deviation is likely to have on the consumer. Existing processes for determining likelihood and severity can be qualitative (Q) based on subjective knowledge of the products and processes, semi-quantitative (SQ) where numbers are assigned to qualitative parameters and fully quantitative (QRA) as in the use of microbiological risk assessment (MRA) methods. Traditionally a decision tree approach is used that, through a set sequence of questions, identifies whether a hazard could occur at unacceptable levels or increase to unacceptable levels. However, this approach requires a manufacturing business to be able to quantify what is deemed acceptable. The use of SQ risk assessment matrices is widespread in the food manufacturing industry.

The World Trade Organisation (WTO) Sanitary and Phytosanitary (SPS) Agreement introduced the term appropriate level of sanitary or phytosanitary protection (ALOP), that is, the level of protection deemed appropriate by a country or Member State establishing an SPS measure to protect human, animal or plant life or health within its borders. The 20th edition of the *Procedural Manual of Codex*

Alimentarius Commission defined a food safety objective (FSO) as the maximum frequency and/or concentration of a hazard in a food at the time of consumption that provides or contributes to the ALOP. This definition recognised that the acceptable level of a hazard may vary at different points in the production, supply and consumption of a food product.

Great care must be applied to the decision process of assessing the severity of a hazard and the likelihood of it occurring, particularly in relation to ensuring all of the requisite information and all of the relevant expertise is available. A decision to incorrectly discard control of a hazard at a specific process step could mean that a substantive CCP is not identified at Principle 2. The Codex publication *Principles and Guidelines for the Conduct of Microbiological Risk Assessment CAC/GL-30 (1999)* provides relevant guidance on this.

3.11   BS EN ISO 22000:2005 states that when selecting control measures, they should be categorised as to whether they are elements of the operational PRP(s), that is, procedural or policy based or elements of the HACCP plan being product or process based. The standard stated that both PRPs and the HACCP plan need to have specific monitoring programmes in place.

Validation is defined in the CAC Guidelines for the Validation of Food Safety Control Measures[1] (CAC/GL 69—2008) as being the obtaining of evidence that a control measure or combination of control measures, if properly implemented, is capable of controlling the hazard to a specified outcome. Validation can also be described as the process of ensuring that the process and procedural controls in place within a manufacturing operation are capable of effectively managing potential food safety hazards should they occur. Therefore, effective validation of control measures is a critical element of the 'due diligence' defence (see 1.6). Validation is an activity undertaken at pre-FSMS design, implementation and as a post-FSMS implementation activity. Ensuring that the design of the FSMS remains valid over time may require re-validation activities to be undertaken. The process of validation is therefore the assimilation and evaluation of data. These data can include, but are not limited to, reference to legislation, scientific data, guidelines, codes of practice or technical information, results from validation studies, historical data arising from monitoring and verification activities or data from similar processes, data from mathematical modelling activities and the use of risk assessment models. Risk assessment models, as previously described in 3.10 can be used to determine whether a specific control measure or a combination of control measures, which may be enforced at different stages of manufacture, is capable of consistently controlling a food safety hazard or reducing a food safety hazard to an acceptable level. Re-validation may be required as a result of system or product failure, process or procedural changes, new scien-

---

[1]For the Guidelines for the Validation of Food Safety Control Measures CAC/GL 69-2008, see http://www.codexalimentarius.net/web/more_info.jsp?id_sta=11022.

tific or regulatory information or evidence of emergent hazards previously unrecognised in the food industry.

The Food Standards Agency (FSA) publication *E. coli O157—*Control of cross-contamination: Guidance for food business operators and enforcement authorities (2011[2]) stresses the importance of not only validating the HACCP plan, but also focusing on validating the control measures in place to ensure bacterial loading on fresh produce is reduced on receipt, that physical separation of materials is effective and that disinfectants are purchased and used in compliance with validated dilution levels and contact times.

3.12   CCPs should be determined where control is necessary to eliminate or reduce the risk of an unacceptable FSO. Account should be taken of the intended circumstances of use by the customer or consumer. This should include both normal intended use and realistic deviations from this. Intended use could include temperature-controlled storage, cooking or reheating of the food product. Measurable critical limits need to be established at each CCP. They are values that separate acceptability from unacceptability in terms of food safety. Target levels and tolerances may also be set, which take into consideration the potential fluctuations within the process and/or provide opportunity to take action before the product is deemed unacceptable (unsafe and therefore rejected).

3.13   At each CCP, a monitoring system must be developed, and appropriate corrective action needs to be determined in the instance that control is lost at a CCP and a target level or critical limit is exceeded. The activities undertaken in monitoring a CCP should be clearly identified in specific work instructions, or similar equivalent. Personnel working at CCPs should be able to demonstrate their appropriate level of competence. The training undertaken and the formal assessment of competence should be recorded. The corrective actions determined must be capable of bringing both the product and the process back under control, where possible before unsafe food is produced. These actions must ensure that any product or material that may have been produced while the CCP was not in control is suitably identified, controlled and adequately assessed to determine appropriate disposition (see 11.9 and Chapter 20).

3.14   Records need to be maintained at each CCP to demonstrate that measurements were undertaken on a routine basis to ensure that CCPs are under control. In the event of a loss of control at a CCP, the resultant actions taken also need to be recorded. These records form part of the manufacturer's due diligence defence and should demonstrate that only competent personnel have been engaged in CCP monitoring activities. Where records are in electronic form, suitable evidence should be available as to how the checks have been undertaken and how the records have been verified.

[2]http://www.food.gov.uk/multimedia/pdfs/ecoli-control-cross-contam.pdf.

3.15   Verification is the activities undertaken in addition to monitoring to determine if the HACCP system is capable of delivering safe food, whether the manufacturing operation is in compliance with the HACCP plan and/or whether the HACCP plan needs modification and review. The HACCP plan should be audited and reviewed at least annually to ensure continuing suitability. CCP records should be verified at intervals defined by the manufacturer to ensure that the HACCP system is implemented and effective. Verification should be undertaken by different personnel to those who undertake the monitoring activities prescribed in the PRP and/or the HACCP plan. It is important that verification activities should not just address the HACCP plan, but also PRPs and their continued effectiveness. The key tool for verification is the audit (see Chapter 7). The results of internal audits, complaint data, product withdrawal and recall data, and data on service levels as well as internal records of rework or rejection and microbiological and chemical analysis should also be considered in this process. Any trends should be identified especially where they indicate a loss of control that has not been suitably managed, and a corrective action programme must be implemented. The frequency of verification should be based on risk assessment.

3.16   The FSMS should be reviewed at least annually. In the event of changes to the product (including formulation and recipes), procedures, processes, responsibilities of personnel, supply or composition of raw materials, packaging and ingredients, consumer use, packaging, storage or distribution activities, or any other factor deemed necessary, a review should be undertaken. In the event of nonconformity as described in 3.14, a review should also be undertaken. Depending on the characteristics of the product, there is the potential for a new, emergent food safety hazard to occur. In this circumstance, a full review should be undertaken, and this may require a reconsideration of all aspects to ensure that the FSMS is still capable of consistently delivering safe products.

Hazard Analysis   3.17   HACCP is just one of a number of recognised methods of hazard
and Operability   analysis including failure mode and effects analysis (FMEA) and
Study (HAZOP)   HAZOP. FMEA seeks to identify which failures in an electrical, mechanical or manufacturing process or system can lead to undesirable situations and the means of detection, safeguards that can be implemented and the required actions. HAZOP is a systematic structured approach questioning the sequential stages of a proposed operation/manufacturing process in order to optimise the efficiency and the management of risk. Thus, the application of HAZOP to the design of a proposed food-related operation should result in a system in which as many CCPs as possible have been eliminated, making HACCP during subsequent operations much easier to carry out.

3.18   HAZOP was developed in the 1960s and was a precursor for the development of HACCP as a means of hazard analysis. It can be used to assess both proposed and existing processes and modifications to current designs. HAZOP is therefore used to identify both food safety hazards and potential operational issues that could lead to food safety or environmental hazards or impact on manufacturing efficiency. The HAZOP approach uses guide words and parameters and identifies potential deviations that could lead to problems such as contamination, filter blockage, corners, bends and dead spaces (which might prove difficult to clean effectively), seal or gasket failure, corrosion or stress fractures.

4   FOOD ALLERGENS

Principle

Great care should be taken (a) to formulate foods so as to avoid, wherever possible, inclusion of major serious allergens as ingredients; (b) to provide appropriate warning, to potential purchasers, of the presence of a major serious allergen in a product; and (c) to organise production, production schedules and cleaning procedures so as to prevent cross-contamination of products by 'foreign' allergens.

Food Allergens

4.1   The problem of food allergens is part of a wider problem, that of all kinds of adverse reactions to foods, which can also result from microbial and chemical food poisoning, psychological aversions, behaviour change and specific non-allergenic responses. In total, over 170 foods have been documented in the scientific literature as causing allergic reactions. Effective management of serious food allergens is an essential part of good manufacturing practice (GMP). The UK Food Standards Agency (FSA) issued draft guidelines entitled 'Guidance on Allergen Management and Consumer Information' Best Practice Guidance on Controlling Food Allergens with Particular Reference to Avoiding Cross-Contamination and Using Appropriate Advisory Labelling (e.g. 'May Contain' Labelling).[1]

4.2   This is an extremely complex problem to which there are no cheap or easy solutions. There are few foods or food ingredients to which someone, somewhere, is not allergic or intolerant, in some cases in very small (microgram) quantities. Allergic reactions may range from relatively short-lived discomfort to anaphylactic shock and death, but all should be treated seriously and safeguarded against by the manufacturer. While not detracting from the responsibility of sufferers (and their medical advisers) to identify the particular foods or food substances to which they are allergic, there is a need for due diligence by manufacturers in considering the use of known allergens as ingredients, in warning the customer or consumer of the presence or potential presence of such allergens, and in preventing accidental cross-contamination of products by allergens used in other products. This is not only a duty of care and a due diligence requirement, but an essential means of minimising the risk of being subject to a product liability claim. While these guidelines should prove useful in providing essential signposts towards developing company GMP in this area, the new development of such a policy requires commitment in the company 'culture', an allocation of funds and resources and a concentrated and sustained effort by everyone led by senior management, and its application and maintenance thereafter.

[1]http://www.food.gov.uk/multimedia/pdfs/maycontainguide.pdf.

Food & Drink – Good Manufacturing Practice: A Guide to its Responsible Management, Sixth Edition.
The Institute of Food Science & Technology Trust Fund.
© 2013 John Wiley & Sons, Ltd. Published 2013 by John Wiley & Sons, Ltd.

23

4.3 Existing or proposed new product formulations should be carefully examined at the product development stage to see whether there is a possibility of excluding food allergens. Of course, in many cases a food allergen is essential to characterise the food, and in such cases label warning must suffice. In some cases, however, where food allergens are present as non-characterising or minor ingredients, it may be possible to effect a substitution. Likewise, a similar approach should be made to food allergens in compound ingredients.

4.4 The accidental presence of a food allergen in a product may arise in three main ways: by accidental mis-formulation, poor labelling control leading to the product containing an allergen being packed or labelled incorrectly, or cross-contamination by a food allergen from a different product.

4.5 Mis-formulation resulting in the inclusion of a food allergen (or any other ingredient) not in the product formulation should be prevented by proper attention to the provisions of Chapter 6. Labelling control procedures are addressed in Chapter 27.

4.6 Cross-contamination of a product by a food allergen from a different product may arise as a result of a variety of activities but could be due to residues in shared equipment, airborne dust or the improper incorporation of rework material without consideration of the allergen problem. The particulate nature of the allergen should also be considered in any allergen risk assessment as this will affect the potential for contamination and the area over which contamination could occur; that is, is the allergen, or allergen-carrying food, a liquid, powder, solid layer or particulate? It should also be considered whether the allergen is of a homogenous distribution through the food or heterogeneous as this will affect the accuracy and repeatability of sampling and monitoring activities. The allergen risk assessment must integrate with the hazard analysis critical control point (HACCP) plan and the food safety management system (FSMS) in its entirety.

4.7 In companies producing on more than one site, or in different buildings on the same site, serious consideration should be given to production segregation in separate buildings. Where separate buildings are not available, separate production equipment or timing separation is recommended. Where production equipment is shared between one or more food allergen-free products and a food allergen-containing product and this is unavoidable, the food allergen-containing product should be run as the last production of the day, immediately before cleaning (e.g. on a shared production line for mixed breakfast cereals, one of which contains nuts, the product containing nuts should be run last). However, it should be recognised that cleaning afterwards, especially in a plant producing dry products will not necessarily guarantee against small quantities of trapped material waiting to be 'carried over' into the first product to go through, and segregation may be the only acceptable solution. The same applies to small quantities of food allergen in airborne dust. In some instances, air handling systems might need to be considered.

4.8   As an example of the measures outlined in 4.7, a formal documented allergen control risk assessment should be implemented on sites where peanuts or nuts, or any other allergen in fact, are processed or stored to ensure that products containing them do not contaminate peanut-free or nut-free product. Nuts should be stored in a designated area and processed on designated lines. Time separation should be considered between nut and nut-free products if they are processed on the same line and also the cleaning procedures, which need to be undertaken after processing nut products (see 4.7). For example, a potential source of contamination could be a spillage of nut products onto the packaging of nut-free products. This contaminated packaging might then cause contamination of nut-free products. If a spillage of nuts occurs, this must be cleared up with care including designated cleaning equipment and disposable wear, or other suitable means, for personnel to ensure that their protective clothing is prevented from contamination. Personnel must ensure that all areas surrounding the spillage are checked for nut debris. All nut debris must be taken immediately to the appropriate disposal area. Any affected goods should be quarantined and disposed of. While nuts have been used in this example, the allergen control procedure should address all likely allergens, or allergenic material that could be present in the food manufacturing premises. The allergen control procedures of suppliers should also be considered so that the likelihood of contamination of raw materials, ingredients or packaging is minimised. Allergenic materials may also arise from processing aids, maintenance and repair activities and factory product trials, and controls should be in place to address all these potential sources. The allergen control risk assessment must also consider the likelihood of contamination and therefore the risk of staff, visitors or contractors bringing allergen-containing food on-site for their own consumption and/or the potential impact of catering functions on-site.

4.9   The incorporation of rework material in a product is covered in Chapter 21, and its provisions should be operated in order to exclude from any food product not containing certain allergens, rework material that contains any of those allergens.

4.10   Appropriate warnings to the potential purchaser are necessary, which involves labelling. Distinctive labelling cannot encompass every one of the 170+ foods documented as causing allergic reactions (or most food would have to carry a warning, and all distinctiveness would be lost). Nor should labelling be regarded as obviating the responsibilities of sufferers (and their medical advisers) to identify the particular foods or food substances to which they are allergic, or the responsibilities of manufacturers referred to in the preceding paragraphs.

4.11   In the **European Union, Directive 2003/89/EC (amending 2000/ 13/EC)** addressed the indication of the ingredients present in pre-packed foodstuffs, while **2006/142/EC** required mandatory label indication of the presence of allergens. These have been superseded by **EU Regulation (EU) No 1169/2011** on the provision of food information to consumers, which came into force in all Member States on 13 December 2011.[2] It will be mandatory on 13 December 2014 (with the exception of Article 9.1, which shall apply on 13 December 2016, and Part B of Annex VI, which shall apply on 1 January 2014). Labels complying with it before those dates will be legal. Annex II list the allergens that must be declared and some exemptions, as follows:

(1) cereals containing gluten, namely, wheat, rye, barley, oats, spelt, kamut or their hybridised strains, and products thereof, except:
   (a) wheat-based glucose syrups including dextrose,[3]
   (b) wheat-based maltodextrins,
   (c) glucose syrups based on barley,
   (d) cereals used for making alcoholic distillates including ethyl alcohol of agricultural origin;
(2) crustaceans and products thereof;
(3) eggs and products thereof;
(4) fish and products thereof, except:
   (a) fish gelatine used as carrier for vitamin or carotenoid preparations,
   (b) fish gelatine or Isinglass used as fining agent in beer and wine;
(5) peanuts and products thereof;
(6) soybeans and products thereof, except:
   (a) fully refined soybean oil and fat,
   (b) natural mixed tocopherols (E306), natural D-alpha tocopherol, natural alpha tocopherol acetate and natural D-alpha tocopherol succinate from soybean sources,
   (c) vegetable oils derived from phytosterols and phytosterol esters from soybean sources,
   (d) plant stanol ester produced from vegetable oil sterols from soybean sources;
(7) milk and products thereof (including lactose), except:
   (a) whey used for making alcoholic distillates including ethyl alcohol of agricultural origin,
   (b) lactitol;
(8) nuts, namely, almonds (*Amygdalus communis* L.), hazelnuts (*Corylus avellana*), walnuts (*Juglans regia*), cashews (*Anacardium occidentale*), pecan nuts (*Carya illinoinensis* (Wangenh.) K. Koch), Brazil nuts (*Bertholletia excelsa*), pistachio nuts (*Pistacia vera*), macadamia or Queensland nuts (*Macadamia ternifolia*) and products thereof, except for nuts

---

[2] http://eur-lex.europa.eu/LexUriServ/LexUriServ.do?uri=OJ:L:2011:304:0018:0063:EN:PDF.
[3] And the products thereof, in so far as the process that they have undergone is not likely to increase the level of allergenicity assessed by the authority for the relevant product from which they originated.

used for making alcoholic distillates including ethyl alcohol of agricultural origin;

(9) celery and products thereof;

(10) mustard and products thereof;

(11) sesame seeds and products thereof;

(12) sulphur dioxide and sulphites at concentrations of more than $10\,mg/kg$ or $10\,mg/L$ in terms of the total SO_2, which are to be calculated for products as proposed ready for consumption or as reconstituted according to the instructions of the manufacturers;

(13) lupin and products thereof; and

(14) molluscs and products thereof.

This list is subject to future amendment. It does not preclude the manufacturer from drawing attention on the label to other food allergens. Article 21, Labelling of certain substances or products causing allergies or intolerances, states that:

(1) Without prejudice to the rules adopted under Article 44(2), the particulars referred to in point (c) of Article 9(1) shall meet the following requirements:

(a) they shall be indicated in the list of ingredients in accordance with the rules laid down in Article 18(1), with a clear reference to the name of the substance or product as listed in Annex II; and

(b) the name of the substance or product as listed in Annex II shall be emphasised through a typeset that clearly distinguishes it from the rest of the list of ingredients, for example by means of the font, style or background colour. In the absence of a list of ingredients, the indication of the particulars referred to in point (c) of Article 9(1) shall comprise the word 'contains' followed by the name of the substance or product as listed in Annex II. Where several ingredients or processing aids of a food originate from a single substance or product listed in Annex II, the labelling shall make it clear for each ingredient or processing aid concerned. The indication of the particulars referred to in point (c) of Article 9(1) shall not be required in cases where the name of the food clearly refers to the substance or product concerned.

(2) In order to ensure better information for consumers and to take account of the most recent scientific progress and technical knowledge, the Commission shall systematically re-examine and, where necessary, update the list in Annex II by means of delegated acts, in accordance with Article 51.

4.12 Labelling deficiencies resulting in allergic reactions may arise (a) because a known food allergen is not recognised by its designation in an ingredients list, for example, declared 'vegetable oil' may be peanut oil; again, few, if any, consumers, knowing they are allergic to milk protein would realise the significance of 'calcium caseinate' in an ingredients list; and (b) because the small print of some long ingredients lists is not conducive to finding specific ingredients to which one is allergic.

4.13 From the viewpoint of food safety, there is a clear need to provide some kind of label warning regarding the presence of any of the food allergens. In some obvious cases, various manufacturers do give voluntary warning, both as a measure of public safety and as a measure of self-protection.

4.14 Inclusion of the name of a food allergen in an ingredients list should not be regarded as adequate warning. The presence, or potential presence, of a food allergen should be separately stated, in a prominent and easily legible way, where it will clearly be seen by a potential purchaser under normal conditions of display. In the UK the FSA Guidance on Allergen and Miscellaneous Labelling Provisions (March 2011) provides more detailed information on food labelling.

4.15 Where a product contains one or more food allergens (whether as individual ingredient(s) or as component(s) in a compound ingredient), the presence of the food allergen should be stated in accordance with 4.13 (e.g. '**Contains PEANUT**'). The terminology should be clearly understandable by the lay person. Thus where calcium caseinate is the food allergen concerned, the warning should read '**Contains MILK PROTEIN**'.

4.16 Where a product nominally free from food allergen is produced on a production line shared with a food allergen-containing product, a suitable warning might be, for example, '**May contain traces of PEANUT**'. However, the use of 'may contain' should not be used as a way of avoiding the measures set out in Sections 4.7–4.10.

4.17 Where a product nominally free from food allergens is produced in the same factory building as a food allergen-containing product, a suitable warning might be, for example, '**Produced in a factory where PEANUT is also handled**'.

Again, this should not be used as a way of avoiding the measures set out in Sections 4.7–4.10.

4.18 Some products may be identified as 'allergen-free' products such as 'gluten free'. In the UK, the FSA's Guidance on the Composition and Labelling of Foodstuffs Suitable for People Intolerant to Gluten (May 2011) has been published, and it provides guidance on the **Foodstuffs Suitable for People Intolerant to Gluten (England) Regulations 2010** and as such identical regulations in Wales, Scotland and Northern Ireland. This legislation implements **EC Commission Regulation No 41/2009**.

4.19 The UK FSA has highlighted six food colours that have been associated with intolerance and hyperactivity in young children. These are sunset yellow FCF (E110), quinoline yellow (E104), carmoisine

(E122), allura red (E129), tartrazine (E102) and ponceau 4R (E124). A voluntary ban has been introduced, and manufacturers are being asked where possible to reformulate their products to remove these colourants. In respect of these six colours, the EU Regulation 1333/2008 on food additives requires that, as of 20 July 2010, a warning is contained on the packaging, that they may have an adverse effect on activity and attention in children.

5 FOREIGN BODY CONTROLS

Principle

The protection of food against contamination with foreign bodies requires the use of hazard analysis critical control point (HACCP) to identify potential sources, with assessment of the types of foreign bodies associated with them and their degree of seriousness. It is important to determine if the foreign bodies are intrinsic, that is, derived from the product, for example, fruit stones or fish bone, or extrinsic, that is, derived from the environment as the method of control will be different. Preventive methods are progressively applied at various points in the process flow, manufacturing, packaging, storage and distribution chain to minimise the risk of the presence of foreign bodies in the product. While the use of automatic inspection devices (metal detectors, X-ray machines and vision systems) is recommended as appropriate, it must be remembered that none of these devices are capable of detecting all foreign body contaminants. The major emphasis must always be prevention. Foreign body control procedures are a key prerequisite to ensuring good manufacturing practice (GMP).

Sources of Foreign Bodies

5.1   It is convenient for practical control to divide sources of foreign bodies into those external to the manufacturing plant and those within the plant and premises. Incoming materials and their packaging from external sources become potential internal sources immediately as they enter the manufacturing premises.

5.2   External sources are frequently associated with characteristic contaminants such as pest predators on fruits and vegetables or parasites in animals. Similarly, particular methods of production, handling and packaging of incoming materials can give rise to characteristic foreign bodies, for example, metal or plastic tags in carcass meat, stones in root crops or slivers of wood in herbs or tea packed in wooden containers. Incoming materials may arrive in primary and secondary packaging, made from metal, glass, plastic, textile, paper or cardboard, and are often on wooden pallets (tertiary packaging). Risk assessment of a material will identify the potential hazards associated with it and its packaging and the appropriate action necessary to minimise their effects. Preventive measures should start at the source of supply and all raw material specifications should include considerations concerning foreign body control and limitation (see Chapter 3). The types of physical contamination that need to be considered in any food safety risk assessment include glass, ceramic, plastic (hard and soft), wood, metal, paint, paper, cardboard, string, stones, pests and parts of pests, building materials and human-origin foreign bodies.

Food & Drink – Good Manufacturing Practice: A Guide to its Responsible Management, Sixth Edition.
The Institute of Food Science & Technology Trust Fund.
© 2013 John Wiley & Sons, Ltd. Published 2013 by John Wiley & Sons, Ltd.

5.3   Internal sources of foreign bodies include the following:

    (i)   the building and installations (see Chapter 12);
    (ii)   the plant and equipment (see Chapter 12);
    (iii)   surface coatings and finishes (see Chapter 12);
    (iv)   extraneous materials (packaging, cleaning materials and equipment, maintenance, engineering and production tools, spare parts, etc.) (see Chapters 12, 14 and 17);
    (v)   personnel (see Chapter 11);
    (vi)   water supply (see Chapter 13);
    (vii)   pest infestation (see Chapter 15); and
    (viii)   recovered or reworked product (see Chapter 21).

Prevention

5.4   Preventive concepts should be considered in:

    (i)   the design of the plant, equipment and buildings and their maintenance;
    (ii)   the management of non-conforming materials, recovered or reworked product;
    (iii)   personnel training and management;
    (iv)   housekeeping and general hygiene; and
    (v)   processing and packaging.

5.5   The examination and analysis of quality control data and consumer complaints records should be used to monitor the effectiveness of preventive action (see Chapter 19).

5.6   All plant, equipment and buildings should be inspected regularly to ensure that nothing has deteriorated, become dirty or become detached, or is likely to do so, and thereby create a risk of contamination of a product (see Chapter 12). Mobile equipment, for example, forklifts and pallet trucks should be used in designated areas, for example, external and internal, and be specific to high- or low-care areas in order to minimise the potential for cross-contamination between the different areas established. The Food Standards Agency (FSA) publication *E. coli* O157—Control of cross-contamination: Guidance for food business operators and enforcement authorities (2011) focuses on the controls that need to be implemented to prevent cross-contamination.[1]

5.7   Personnel should be instructed and encouraged to report immediately any incident of contamination or potential contamination of people, materials, packaging, equipment or the product.

5.8   Personnel must be issued with suitable protective clothing. Overalls should be knee length, have internal pockets only and non-detachable fastenings. Loose items, unless required to carry out necessary work, should be banned in production areas. These include keys, coins, mobile phones and so on. In the event that loose items, for example,

---

[1]http://www.food.gov.uk/multimedia/pdfs/ecoli-control-cross-contam.pdf.

keys, are required, a 'sign-out and sign back in' control protocol should be in place that also identifies the corrective action required in the event an item is lost. All wristwatches, jewellery and bracelets should be prohibited. Plain wedding rings, or wedding wristbands, and secure 'sleeper' continuous loop earrings are possible exceptions if control procedures have been developed to manage these items effectively within the production area. Adequate head, body and facial hair coverings must be provided and properly used, and should not be kept in place with the aid of hairpins or other fastenings, which could drop off. All head hair should be fully contained in the head covering; sleeves should come to the wrist to prevent the loss of arm hair and potential contamination of the product. Any hoods on personal items of clothing should be under the protective clothing at all times. Clothing with high collars including those that have zips with metal 'fobs', which could dislodge and fall into the product, should be underneath the protective clothing at all times. Exposed cuts and abrasions must be covered by a waterproof dressing, which should be metal detectable, brightly coloured and easily seen against the background of the product. If disposable gloves are used, they should be checked regularly for any signs of damage or loose pieces that could fall off and contaminate the product. Gloves should be of a distinctive colour, be made of non-allergenic material and must not shed fibres or particles. Smoking, eating and drinking and the use of chewing gum should be restricted to designated areas, away from food production and storage areas, with adequate waste disposal and hand-washing facilities being provided. Employee facilities such as lockers, changing areas, toilets, smoking areas and rest and food areas should be designed and constructed so that they can be maintained in a clean and hygienic manner. The use of clear lockers in food production entry areas aids inspection during hygiene audits. The training programme should explain the necessity for the restrictions and disciplines required in production and storage areas (see Chapter 11).

5.9   Good housekeeping (see Chapters 6, 14 and 15) requires clear instructions concerning the use and disposition of general materials to appear in master manufacturing instructions, plant operating instructions, work instructions, maintenance and service instructions and cleaning manuals. Good housekeeping includes the general tidiness and cleanliness of production and storage areas and also covers infestation control. Buildings must be protected against penetration by animals, birds, rodents and insects by adequate maintenance and proofing. Secondary defences such as poisoned baits, flying insect electrocutors, sticky boards and sprays should be used in appropriate areas to deal with animals and insects that penetrate the building, but should not be seen as the primary means of control. It is generally accepted practice to use non-toxic rodent baits internally within a food manufacturing unit unless a pest infestation has been identified. Due care should be taken to ensure that the infestation controls in themselves do not prove a means of food contamination so they must be sited in suitable locations and securely fastened in place. Wood, glass and paint should, where practicable, be eliminated from open food areas and equipment (plant).

5.10   Processing should be designed to include procedures that will minimise the risk of foreign body contamination of the product. Containers may be kept inverted, where practicable, and should be cleaned by jets of filtered air or potable water before being filled. Typical detection and removal activities in a manufacturing process include visual inspection and sorting, air and liquid flotation, spinning, sieving, sifting, washing, filtration, magnets and magnetic grids and plates, metal detection, optical grader/sorter, and X-ray detection and other forms of scanning.

Effective metal detectors should be employed on the production lines and plant at suitable points in the process. More elaborate methods, such as X-ray examination, exist and may be useful. In addition to the use of X-ray examination in factories, cargoes in containers or road vehicles may be X-rayed at some ports during import/export. Where metal detectors are used within the process, they should always include automatic rejection systems and closed containers to hold reject materials. The closed containers should be locked at all times unless being opened by a designated person during monitoring checks. Detection of foreign matter may lead to material that can in some instances be reprocessed. This decision must be undertaken by suitably competent people who understand the full implications of allowing the product to re-enter the food supply chain. Those responsible for the decision should provide adequate instructions to staff and ensure that suitable monitoring activities are in place.

5.11   A formal risk assessment should be undertaken to determine whether foreign body detection equipment should be used within the manufacturing process to detect and/or remove foreign body contamination (see Chapter 3). This risk assessment should be undertaken by competent staff and be formally recorded. The frequency of subsequent review of the risk assessment should also be defined and complied with.

5.12   Methods of foreign body control should be defined in formal documented procedures. These should include instructions for undertaking the foreign body control procedure and actions to be taken in the event that monitoring identifies product, procedural and/or equipment failure. These foreign body control procedures should address the action to be taken to identify the product that may be affected, its location, the protocol for recall back to the production unit and the procedure for re-inspection. All materials and/or products should be re-inspected that have passed through the inspection method or detection equipment since the procedure was last known to be fully operational, that is, working correctly. All foreign body detection and removal measures should be validated and re-validated as required (see Chapter 3). As previously identified, foreign body controls can include specific process steps such as sieving and filtering. An inventory of filters and sieves should be in place that details the unique equipment identification number, location, type, size in millimetres or microns, and the frequency of inspection. Magnets should also be included on the list if applicable. Sieves and filters should be risk assessed to ensure that they themselves do not present a foreign body risk. This assessment should include reviewing if the sieves are made of metal detectable material or are of a contrasting colour to the food. The review should also ensure the equipment is controlled by documented inspection procedures that identify the action to be taken if the sieves or filters are found to be broken or damaged; the maintenance procedures that have been established for the equipment; and the level of training required for operators and those inspecting the equipment. If foreign bodies are found during sieve or filter inspection, then the nature and number of the foreign bodies and the corrective action that has been taken to rectify the problem should be recorded.

5.13   Consideration should also be given to:

- the type of food being analysed especially particle size and the packaging type (if detection is undertaken following packing). A risk assessment should be undertaken if metal detection is deemed appropriate as to the type of detector and its sensitivity to ferrous, non-ferrous and stainless steel metal and the degree of sensitivity required. Product packed into metal, metal laminated or foil packaging should be considered in terms of the alternative metal controls that may be required within the manufacturing process;
- the design of detector required. In-line pipe metal detectors and the conveyor-type systems (suitable for large items), which have a belt stop system in place if metal is detected, should be fitted with an audible or visible alarm. Personnel working in the area need to be trained to understand the reasons why such an alarm should go off and the actions that need to be taken. For smaller items, a conveyor-type metal detection system should be fitted with an automatic reject system that has been validated and is routinely monitored to ensure effectiveness. Rejected product should be transferred automatically into a secure, locked box, and there should be an alarm system in place should metal be detected. Personnel working in the area should be aware of the significance of items being rejected and be able to take appropriate corrective action;
- the appropriate mesh/sieve size of filters or sieves;
- the means of validation and the limit of detection of the equipment and whether this represents an acceptable food safety limit. This information will normally be held in the HACCP plan (see Chapter 3);
- the location, that is, where the detector is to be positioned on the process line and whether this location could be bypassed by product in the event of a specific production activity;
- the frequency of monitoring or inspection of the equipment/ method and the verification activities employed to ensure such measures are effective;
- the mechanism for rejection following failure (reject arm, locked box system, automatic line stop). The level of security at such devices must be determined as well as the authority for overriding

fail-safe mechanisms, if fitted. Changes to settings or the turning off of reject mechanisms should only be authorised by designated personnel and such decisions must be recorded. Rejection procedures should be formalised, and only authorised personnel should be able to access the locked box or product that has been rejected by the foreign body detector;

- the method of detector calibration (either manual or automatic) as well as servicing and maintenance requirements;
- developing documented procedures that define equipment start-up and operating instructions, and the routine monitoring, testing and calibration of detector equipment, including metal detectors;
- the reporting of incidents and the implementation of effective, timely corrective action. The incident and the corrective action taken should be recorded in the event of a failure of the foreign body detector. This will include stopping the process, and the subsequent isolation, quarantine and re-inspection of all items produced since the last acceptable test result.

All materials and/or products should be re-inspected that have passed through the inspection method or detection equipment since the procedure was last known to be fully operational, that is, working correctly. Material detected or removed by the equipment in the case of filters or sieves should be inspected. It should be retained as evidence, or if more appropriate, digital photographs should be taken, so that a full examination can be undertaken and appropriate preventive and corrective action is adopted.

5.14 Where recycled glass containers are used, provision should be made for the inclusion of automated vision inspection systems to inspect the containers for damage and contamination, including cleaning residues.

5.15 The delivery or storage of materials may involve intermediate packaging to prevent damage. This will subsequently have to be removed. This should be designed to minimise the risk of contaminating the product during its removal. For example, pastry materials are sometimes filled hot into boxes lined with a loose plastic film bag. Creases in the bag become surrounded by pastry that traps the film firmly when it cools. This type of packaging should be avoided, where complete removal cannot be assured. In other cases, if used, internal liners or the film should be highly and contrastingly coloured. Packaging should be clean prior to removal of contents. Materials packed in sealed (seamed) metal containers have the obvious hazard of metal swarf being created when they are opened. Cartons should be staple free; paper sacks should be easy-open, string free and not cut with blunt knives. Plant operatives should be trained to open packaging carefully to avoid product contamination, for example, by the misuse of case-opening knives. The use of a deboxing/debagging area with the transfer of ingredients into internal packaging/containers before being transferred to the processing area is recommended to minimise the potential for foreign body contamination. Knife control procedures should be implemented (see 12.43). Other sources of metal to be considered are cutting blades on equipment, chain mail gloves, aprons and arm protection, needles, nuts and bolts, and wires. Control procedures should be in place to monitor metal items in the production area that can become loose and/or damaged, and appropriate corrective action procedures should be in place in the event of such loss or damage.

5.16   All final packaging used by the manufacturer for products should be examined to ensure compliance with the specification against which it is purchased. In addition to this examination, the detailed appraisal of a manufacturing-scale sample, such as a pallet load on a production line, is strongly recommended. This allows the performance of a high-risk packaging material such as glass to be better assessed before acceptance of the bulk delivery for use. In order to implement this protocol, effective lot traceability of packaging material is required. Packaging materials should be brought to their points of usage in minimal quantities (see Chapter 6).

5.17   Brittle material control procedures to control glass, hard plastic or ceramics should be developed (see Chapter 12, Sections 12.36–12.42).

5.18   When product containers are cleaned before use, filtered air or potable water should be used. The effectiveness of these cleaning procedures should be routinely monitored and verified and appropriate action taken based on the results. Where equipment is used to transport, store or hold the product, then the cleaning of these items should be monitored to ensure suitable efficacy. A reject system should be in place to prevent the use of dirty containers that come into direct contact with the product and therefore could present a foreign body risk.

5.19   Where contamination occurs intermittently or infrequently, either systematically or randomly, no practical sampling scheme is likely to detect the fault. Analysis of data produced as a result of monitoring, verification and consumer complaints will indicate any pattern of foreign body contamination and any changes can be subjected to trend analysis that may show the significance or otherwise of the changes (see Chapter 19).

5.20   A risk assessment should be undertaken of offices that open directly into production/storage areas or workstations within production areas. Office equipment, chairs, tables and desks should not be made of wood and should be constructed so that they are easy to clean. Pencils should not be used. Personal items should be kept to a minimum, and no eating or drinking should take place unless water is consumed following an appropriate risk assessment having been made. Stationery items such as paper clips, staples, pens and so on all present a potential foreign body hazard and should be adequately controlled. Quality control workstations should also be risk assessed for the potential for equipment to become lost or damaged and then present a foreign body risk to the product.

5.21 Regrettably, contamination of products by foreign bodies may on occasion be caused deliberately:

(a) during production by an unstable, malicious or disgruntled person;
(b) somewhere in the distribution/retailing chain, by an individual seeking to harm or blackmail a company; or
(c) after purchase, by an individual seeking financial gain or publicity.

While it is difficult to establish complete safeguards against case (a), it is less likely where good industrial relations are fostered. In addition, management should carefully weigh the dangers of allowing particular persons under notice of dismissal or redundancy to work out the period of their notice.

A documented risk assessment should be undertaken to determine the security procedures required in view of the site layout, nature of the products produced and the processing employed and the specific area of the site. It may be prudent to identify high-risk (HR) and low-risk (LR) areas with regard to product security and have additional security controls in place in HR areas. This can include (see 12.2), but is not limited to, colour of protective clothing for HR areas; restricted entry procedures for authorised personnel only including fingerprint entry to HR areas; and site security arrangements as a whole in terms of fencing, security staff and closed-circuit television (CCTV). A visitor and contractor reporting system should be in place, and the person responsible for such individuals is responsible for monitoring their activities when they are on-site. The use of photographic and recording equipment including mobile equipment that contains these functions must be strictly controlled. All such equipment brought onto the production site must be authorised by the site management, for example, memory sticks, laptops, IPods, IPads and mobile phones. Staff should be encouraged to report unknown individuals when they see them on-site. The security risk assessment should be reviewed a minimum of annually and more often in the event of a breach of security arrangements.

Case (b) hazards should be minimised wherever practicable by the use of tamper-evident packaging and tamper-evident seals. As regards (c), careful study of the relevant facts and laboratory examination of the foreign body should be carried out, the results of which may sometimes demonstrate the probability (or even certainty) that it had been introduced subsequent to the pack having been opened.

6   MANUFACTURING ACTIVITIES

Principle

The operations and processes used in manufacture should, with the premises, equipment, materials, personnel and services provided, be capable of consistently producing finished products, which conform to their specifications and are suitably protected against contamination or deterioration. Defined manufacturing procedures that address manufacturing operations and associated activities and necessary precautions, are required to ensure that all individuals concerned understand what has to be done, how it is to be done, who is responsible and how to avoid mistakes, which could affect food safety or quality. For each product, this is provided in the master manufacturing instructions (see Chapter 9).

General   6.1   This chapter deals with those aspects of manufacturing operations and activities that are of general application to any food or drink manufacture. There are, of course, particular additional points and problems concerned with manufacture involving specific types of processing, materials or products. These are the subjects of the chapters in Part II.

Process Evaluation   6.2   Before the introduction of master manufacturing instructions for a product, trials should be carried out to establish whether the formulation, and the proposed methods and procedures specified therein are suitable for factory production, and are capable of consistently yielding product that complies with the finished product specification. Validation of food safety process parameters and the effectiveness of prerequisite programmes (PRPs) should also be undertaken at this point and as necessary further amendments and trials should be completed until these conditions are fully satisfied (see Chapters 3 and 8).

     6.3   Similar evaluation, and re-validation, should be carried out in connection with any significant proposed change of raw material, equipment or production and/or inspection methods (see Chapters 3 and 8).

     6.4   Similar evaluation should be carried out periodically, to check that the master manufacturing instructions are being followed, that they still represent an effective and acceptable way of achieving the specified product and that they are still capable of consistently doing so (see Chapter 7).

Production   6.5   Adequate resources should be provided in the way of premises, equipment, materials, suitably trained personnel, services, information and documentation, in each case of appropriate quantity and quality (see Chapters 2, 9–12, 14, 16, 18 and 26) to enable the requisite quantity and quality of finished products to be produced.

Food & Drink – Good Manufacturing Practice: A Guide to its Responsible Management, Sixth Edition.
The Institute of Food Science & Technology Trust Fund.
© 2013 John Wiley & Sons, Ltd. Published 2013 by John Wiley & Sons, Ltd.

	6.6	Production should be carried out in full compliance with the master manufacturing instructions from which no departure should be permitted except by written instructions from the production manager and the quality control manager, indicating the nature and duration of the departure, and agreed and signed by them. This is often termed 'a concession'.
	6.7	Incentive bonus schemes can create potential hazards and, viewed from the standpoint of food safety and quality, are best avoided. If, however, the provision of an incentive bonus scheme is company policy, it should be so designed as to discourage operators from taking unauthorised 'short cuts', for example, by building into the formula for bonus calculation a 'quality factor' and/or penalty for observed deviation. In general, prevention of unauthorised short cuts is primarily a task for management through supervision. Where operators have ideas for process improvement, they should be encouraged to raise them (e.g. through suggestions schemes) so that they may be properly evaluated.
Operator Procedures	6.8	Operating instructions for production operators should be written in clear, unambiguous, instructional form, and should form a key part of operator training. Due regard should be given to the reading or language difficulties of some operators.
	6.9	Particular attention should be paid to problems that may arise in the event of stoppages, breakdowns or emergencies, and written instructions should be provided for the corrective action to be taken in each such case.
Raw Materials	6.10	Each raw material should comply with its written specification.
	6.11	Each delivery or batch should be given a reference code to identify it in storage and processing, and the documentation should be such that, if necessary, any batch (see definition) of finished product can be traced back to the deliveries of the respective raw materials used in its manufacture and correlated with the corresponding laboratory and/or supplier records. Deliveries should be stored and marked in such a way that their identities do not become lost. Procedures should comply with **EC Regulation No 178/2002 of the European Parliament and the Council of 28 January 2002 laying down the general principles and requirements of food law, establishing the European Food Standards Authority and laying down procedures in matters of food safety (OJ L 031, 01/02/2002 P. 001 – 0024) and Article 11, Traceability, which came into mandatory force on 1 January 2005.**[1] Guidance on the general principles and basic requirements for traceability system design and implementation is contained in ISO 22005:2007 (see Chapter 10).
	6.12	Deliveries of raw materials should be quarantined until inspected, sampled and tested in accordance with the quality management

[1]http://europa.eu.int/eur-lex/pri/en/oj/dat/2002/1_031/ 1_0312002 0201en00010024.pdf.

system (QMS) and Chapters 26 and 29, and released for use only on authority of the quality control manager, taking account of any certificate of analysis or conformity accompanying a delivery. Particular care should be taken where a delivery of containers appears from markings to include more than one batch of the supplier's production, or where the delivery is of containers repacked by a merchant or broker from a bulk supply. Where appropriate, immediate checks should be carried out for off-flavours, off-odours or taints, and particularly in the case of additives, testing should include test of identity, that is, establishing that the substance is what it is purported to be. (N.B. In a multi-container delivery, it is impracticable to check the identity of the contents of every container on arrival, but operators should be trained and encouraged to report immediately anything unusual about the contents when a fresh container is brought into use.)

6.13   Temporarily quarantined material should be located and/or marked in such a way as to avoid risk of its being used accidentally. Material found to require pretreatment before being acceptable for use should be suitably marked and remain quarantined until pretreatment. Material found totally unfit for use should be suitably marked and physically segregated pending appropriate disposal.

6.14   In the case of a bulk delivery by tanker, preliminary quality assessment should be made before discharge into storage is permitted.

6.15   All raw materials should be stored under hygienic conditions, and in specific conditions (e.g. of temperature, relative humidity) appropriate to their respective requirements as indicated in their specifications, and, **with due regard to the requirements of legislation on substances hazardous to health, for example, in the UK, the Control of Substances Hazardous to Health (COSHH) legislation**.

6.16   Stocks of raw materials in store should be inspected regularly and sampled/tested where appropriate, to ensure that they remain in acceptable condition.

6.17   When issuing raw materials from store for production use, correct stock rotation should normally be observed, unless otherwise authorised or specified by the quality control manager. This process is very often termed 'FIFO', that is, first-in-first-out stock rotation.

6.18   A formal procedure and associated documentation should be established and followed for the issue of raw materials from store. The personnel responsible for issuing raw materials should be formally defined.

6.19   When a raw material has been issued but not used as planned (e.g. because of a plant stoppage), the quality control manager or designate should advise as to its disposition. This is critical where items require specific storage conditions such as temperature control in order to maintain food safety or quality.

6.20   Depending on the product being manufactured, the ingredients involved and the nature of the process and equipment, dispensing of the required quantities of ingredients could take various forms, including manual dispensing by weight or volume, automatic dispensing of batch quantities by weight or volume, or continuous metering by volume; the form(s) actually taken will be stated within the master manufacturing instructions. In each case, the weighing and/or measuring equipment should have the capacity, accuracy and precision appropriate to the purpose, and the accuracy should be regularly checked (see Chapter 25).

6.21   Where batch quantities of a raw material have to be dispensed manually into containers in advance, this should be done in a segregated area. Where manual pre-dispensing of relatively small and accurate quantities (e.g. additives) is required, this should be done by, or under direct supervision of, competent staff.

6.22   Records should be kept to enable the quantities of materials issued to be checked against the quantity or number of batches of product manufactured.

6.23   Where an operator controls the addition of one or more raw material to a batch, the addition of each ingredient should be recorded at the time on a batch manufacturing record to minimise risk of accidental omission or double addition. Ingredients with key health and safety implications may need to be formally signed off before addition to the batch, for example, allergens or preservatives.

Packaging Materials   6.24   Each packaging material should comply with its specification (including any legal requirements). The specification should be such as to ensure that:

(a) the product is adequately protected during its expected life under normally expected conditions (with a safety margin for adverse storage);
(b) in the instance of packaging coming into immediate contact with the product, there is no significant adverse interaction between product and packaging material. In the European Union (EU), this includes compliance with the **EU Framework Regulation (EC) No 1935/2004,**[2] which defines general requirements for all food contact materials, and a number of specific directives that cover single groups of materials and articles listed in the framework regulation, as well as directives on individual substances, or groups of substances used in the manufacture of materials and articles intended for food contact[3];
(c) where the packaged product undergoes subsequent treatment, whether by the manufacturer, caterer or consumer, the packaging will adequately stand up to the processing conditions and no adverse packaging/product interaction occurs;

[2]http://europa.eu.int/comm/food/food/chemicalsafety/foodcontact/framework_en.htm.
[3]http://www.europa.eu.int/comm/food/food/chemicalsafety/foodcontact/eu_legisl_en.htm.

(d) the packaging is capable of providing the necessary characteristics and integrity where the preservation of the product depends on the pack; and

(e) the finished pack will carry the statutory and other specified information in the required form and location.

6.25   Where packaging material (e.g. labels, printed packages, lithographed cans) carries information required by law, the quality control manager should ensure that the specification is updated as required to comply with new legal provisions. In some organisations this responsibility may be within the product development department. The quality control manager should ensure that stocks of packaging materials that no longer comply are quarantined for modifications (if possible and desired) or destruction. Similar provisions and precautions should apply in the case of contract packing for a customer where the latter requires changes to other label information (see Chapter 24).

6.26   Each delivery or batch of packaging should be given a reference code to identify it in storage and processing, and the documentation should be such that, if necessary, any batch of finished product can be traced to deliveries of the respective packaging materials used in its manufacture and correlated with the corresponding laboratory records. Deliveries should be stored and marked in such a way that their identities do not become lost (see Chapter 10).

6.27   Deliveries of packaging material should be quarantined until inspected, sampled and tested in accordance with the QMS and Chapters 26 and 29, and released for use only on authority of the quality control manager. Deliveries of packaging should be adequately inspected to confirm that they comply with the relevant specification and that the external packaging is intact and unopened. Damaged, previously opened or dirty cartons of packaging must not be accepted due to the risk of contamination. Operators should be trained and encouraged to report immediately anything unusual about the appearance, odour or behaviour of packaging materials issued especially colour differences, changes in print quality or problems with sealing.

6.28   Temporarily quarantined packaging material should be located and/ or marked in such a way as to avoid risk of its being accidentally used before release. The personnel responsible for the decision on its disposition should be formally defined. Material found totally unfit for use in packaging operations should be suitably marked and physically segregated pending appropriate disposal.

6.29   All packaging materials should be stored in hygienic conditions and as indicated in their respective specifications.

6.30   Stocks of packaging materials in store should be inspected regularly to ensure that they remain in acceptable condition, and appropriate infestation/pest control procedures must be in place for storage areas (see Chapter 15).

6.31 In issuing packaging material from store for production use, stock rotation should normally be observed, unless otherwise authorised or specified by the quality control manager. This process is very often termed FIFO stock rotation.

6.32 A formal procedure and associated documentation should be established and followed for the issue from, and the return of, part-used batches of packaging to store. The procedure should address the need to reseal part-used boxes of packaging to prevent foreign body contamination. The personnel responsible for issuing and reconciling packaging stocks should be formally defined.

6.33 Where a company manufactures more than one product, or more than one version of a single product, the greatest care should be taken to check that the correct packaging is issued for the product to be manufactured, and that no incorrect packaging materials left over from a previous production run of a different product or a different version are left in the production area where they might accidentally be used. This is especially important when products are placed on promotion and the packaging specification is amended, and then after a given duration, the product is packed in the original packaging. Particular attention should be paid to controlling adhesive, flash and price labels, especially where the design of the labels is very similar.

6.34 Under no circumstances should primary food packaging be used for anything other than its intended purpose.

6.35 Where packaging is reference coded, priced or date marked for use, care should be taken to ensure that only material carrying the correct production date is used. Surplus material left from earlier production runs and no longer bearing a valid reference or date should not be left in the production area. Where the reference and/or date is applied during the manufacturing operation, care should be taken to check and ensure that the marking machine is set for the correct reference and date. Where packaging or labels are generated in an off-line print room, procedures need to be in place to sign off/verify that the print details are correct when released from the print room and when accepted by the production line. A protocol should be in place to ensure that any unused packaging or labels are suitably controlled so that they cannot be inadvertently used in production. Label approval, monitoring and verification activities must be formally defined and appropriate training given to those personnel assigned these tasks so they understand the meticulous inspection required and the consequences if an incorrect label is not identified and used within production. Crisis management, product recall and product withdrawal procedures must be in place that address the procedures to be followed in the event that incorrect product labelling has occurred especially if the product is known to, or may potentially contain an allergen that is then not identified on product labelling/packaging.

Processing and Packaging

6.36 Where a company manufactures more than one product or more than one version of a product, and there is more than one production line, production layout should be such that confusion and possible cross-contamination are avoided (see Chapter 5).

6.37 Whether in single-line or multiple-line production, particular care should be taken, in terms of production layout and practices, to avoid cross-contamination of one product by another (see Chapter 5).

6.38 On a production line, in order to avoid confusion, the name and appropriate reference to the product being processed or packaged should be clearly displayed, or otherwise communicated.

6.39 Before production begins, checks should be carried out to ensure that the production area is clean and free from any products, product residues, waste material, raw materials, packaging materials or documents not relevant to the production to be undertaken, and that the correct materials and documents have been issued and the correct machine settings have been made. All plant and equipment should be checked as clean and ready for use. A formal form or checklist may be designed for this purpose.

6.40 Processing should be strictly in accordance with the master manufacturing instructions subject to any variations approved (as in 6.6), and by detailed procedures set out for operators in the plant operating instructions. Master manufacturing instructions, product or process specifications should define the following:

(a) pre-startup, close-down and product changeover checks to ensure the line is cleared for production and closed down correctly at the end of the shift;
(b) the product recipe in terms of ingredients and proportions;
(c) mixing instructions including machinery to be used, duration of mixing and processing requirements;
(d) equipment settings, processing parameters (time, pressure and temperature);
(e) the control of in-process monitoring devices and divert systems if minimum processing parameters are not reached;
(f) filling requirements in terms of volume or weight;
(g) criteria for in-process foreign body inspection or detection;
(h) inspections to ensure processing conditions are consistent throughout the batch being processed;
(i) packaging and labelling requirements especially at start-up and product changeover;
(j) sampling criteria in terms of parameters to be checked and the frequency of sampling; and
(k) additional food safety and quality criteria.

6.41 Process conditions should be monitored and process control carried out by suitable means including, as appropriate, sensory, instrumental and laboratory testing, and on-line checking of correct packaging and date marking. Where checkweighers, filler systems, continuous recorders or recorder/controllers are in use, the data produced should be checked by quality control and charts or printouts retained in either electronic or paper form as process records. This is particularly important at process steps that are deemed by the manufacturer as a critical control point (CCP) for a food safety hazard (see Chapter 3).

6.42 There should be regular and recorded checks by appropriate personnel on the accuracy of all instruments used for monitoring processes (e.g. temperature probes/gauges, pressure gauges, flow meters, checkweighers, colour measurement devices, metal detectors and X-ray machines). The frequency of checks should be established to ensure that instruments are always correctly calibrated, with an accuracy related to national standards (see Chapter 25).

6.43 Effective cleaning of production premises and equipment must be carried out (see Chapter 14).

6.44 All persons working in or visiting the production area must comply with the requirements of personal hygiene, and adequate facilities must be provided, as summarised in Chapter 11.

6.45 General 'good housekeeping' should be practised including prompt removal of waste material, precautions to minimise spillage or breakage, prompt removal and clean-up of any spillage or broken packaging, and the removal of any articles that might enter the product as foreign matter, otherwise known as a 'clean-as-you-go' policy (see Chapter 14).

6.46 Where appropriate, foreign matter detectors should be used (see Chapter 5).

Intermediate Products

6.47 After its preparation, an intermediate product should be held until checked and approved by quality control for compliance with its specification. If required to be stored before further processing, it should be stored as designated in the specification, and suitably reference marked and documented so that it can be traceable and correlated with the lots of raw materials from which it was made and the batch(es) of finished product in which it is subsequently incorporated (see Chapter 10).

6.48 A batch of intermediate product found to be defective should be quarantined pending reworking or recovery of material or outright rejection as the case may be. The personnel responsible for the decision on its disposition should be formally defined (see Chapter 21).

Finished Product

6.49 Packed finished products should be quarantined until checked and approved by quality control for compliance with the appropriate finished product specification.

6.50 An approved batch of finished product should be suitably marked to identify it, and stored under the appropriate conditions (e.g. of temperature or relative humidity) stated in the finished product specification. Under the traceability requirement of **EU Regulation 178/2002**, identification mark must be provided in a traceability document to any retailer to whom part of that batch is sold.

6.51 Where a batch of finished product fails to meet the specification, the reasons for failure should be thoroughly investigated. Defective finished product should remain quarantined pending reworking or recovery of materials or disposal as the case may be. The personnel responsible for the decision on its disposition should be formally defined (see Chapter 21).

Storage

6.52 Desirable features of storage premises are outlined in Chapter 12 and Chapter 18. The following paragraphs refer to some desirable storage practices for raw materials, packaging materials, intermediate products and finished products prior to distribution; they supplement those already referred to in 6.16, 6.29, 6.30 and 6.31. Aspects of storage of goods in the distribution system after they have left the manufacturing unit are dealt with in Chapter 23.

Transport

6.53 Materials or products should be transported within the factory premises in such a way that their identities are not lost; that there is no mixing of materials or products approved for use or despatch with those that are quarantined; that by-products, particularly those not intended for human food use, do not lead to contamination of other materials; that spillage is prevented; no breakage or other physical damage is caused to the goods being transported; and that goods being transported are not left in adverse conditions or otherwise allowed to deteriorate.

Principle   *Management review is the process of reviewing the quality management system (QMS)/food safety management system (FSMS) at regular intervals. The review is undertaken to determine the degree of management control and its effectiveness and verify that the food safety and quality policy and food safety and quality objectives are suitable, and being complied with. It is also an opportunity to identify any areas for improvement. An internal audit programme should support the management review process. It provides an input to management review by measuring the level of conformance with the QMS/FSMS and the effectiveness of the QMS/FSMS in achieving food safety and quality objectives.*

Management Review 7.1   Management review is formally addressed at regular structured management meetings. The organisation of the management review meetings includes determining:

- who will be on the management review team;
- who will chair the meetings;
- how often the meetings will take place;
- the agenda that will be followed; and
- who will take and maintain the minutes of the meeting.

The meeting minutes form part of the food safety and quality records of the organisation (see Chapter 9) and need to be retained to provide objective evidence that the meeting took place and the actions that were agreed as a result of the meeting. Clear actions need to be agreed with responsibilities for action and timescales for completion. The channels of communication of the meeting decisions and the actions that have been agreed must be formalised. The frequency of meetings can be monthly, bimonthly or quarterly. If the meetings are any less frequent, then they become a historic review rather than a real-time management process that drives continuous improvement. For the same reason, minutes need to be circulated to those concerned without delay as a guide to agreed decisions and in order to facilitate prompt action.

Inputs to the Review Process 7.2   Inputs to the management review process can include the following:

- areas of non-compliance including incidents, customer complaints, non-conforming items and status of existing corrective actions as well as the effectiveness of previously completed corrective actions;
- progress in complying with previous management review meeting minutes, and action plans;
- progress in complying with the FSMS and hazard analysis critical control point (HACCP) plan and any non-conformance and subsequent corrective action;

Food & Drink – Good Manufacturing Practice: A Guide to its Responsible Management, Sixth Edition.
The Institute of Food Science & Technology Trust Fund.
© 2013 John Wiley & Sons, Ltd. Published 2013 by John Wiley & Sons, Ltd.

- progress in complying with food safety and quality objectives and trends in food safety and quality costs such as level of rejection, rework, cost of customer complaints;
- preventive actions and opportunities for improvement that have been or need to be addressed;
- results from audits, internal, external and third-party audits;
- changes to the documented system since the last review;
- general customer feedback and analysis of customer and/or consumer complaint data including service levels and subsequent investigations and actions taken;
- changes to the organisational structure or individual job responsibilities;
- analysis of supplier performance and approval of new suppliers and contractors;
- analysis of product and service performance;
- changes that could affect the QMS/FSMS system including new market requirements, legislation, technology, products and processes; and
- training programmes and the current training needs/plans including effectiveness of previous training.

7.3 If an integrated management review process is undertaken, inputs could also include social accountability, animal welfare, corporate social responsibility, personnel health and safety and environmental issues.

Outputs from the Review Process

7.4 Measurable outputs that demonstrate the effectiveness of the management review process include:

- continued compliance with statutory, legislative and market requirements;
- improved allocation of resources;
- continued focusing of the food safety and quality policy and food safety and quality objectives to the issues affecting the organisation;
- improved planning and communication on future changes within the business including new products, technology and processes, new suppliers, changes to the organisational structure and individual job responsibilities or the documented system;
- improved management of the corrective action programme;
- monitoring of supplier performance and improved communication with regard to issues of service and raw material/ingredient performance;
- improved relationships with customers;
- measurement of food safety and quality costs on an ongoing reduction in non-conforming product and service; and
- continued focusing of training programmes and identification of current training needs.

Resourcing the QMS/FSMS

7.5 It is the role of the senior management team, using the management review process as a driver, to provide the resources required to fulfil food safety and quality objectives and to ensure legal compliance.

Management review needs to consider whether existing human, physical and financial resources are adequate to achieve this and, where this is not the case, drive the implementation of appropriate corrective action so that adequate resources are in place.

Legislative and Industry Guidance   7.6   The senior management review process should have a formal process in place to ensure that the organisation can access and then address within its management systems and procedures scientific and technical developments including the emergence of new food safety hazards or new control measures, industry guidance and codes of practice, and changes to legislation both in the country of manufacture and those countries to which the organisation is seeking to export its products.

Internal Audit   7.7   An auditing programme needs to be drawn up and include all activities within the scope of the QMS/FSMS. The frequency of audits should be established by risk assessment. Auditing is the management tool that identifies whether the quality manual, associated procedures and systems, including prerequisite programmes (PRPs), have been developed, implemented, are effective and are being complied with. The audit process will also identify whether the QMS/FSMS meets a system standard such as the British Retail Consortium (BRC) Global Standard for Food Safety, alternative private or retailer standard, ISO 22000 or EN ISO 9001:2000. Internal, or first-party, audits are where the organisation develops an auditing programme to audit itself in terms of products, procedures and processes. The resource required for the internal auditing programme should be reviewed including the required number of auditors and the training required in order to develop the auditors' specific auditing skills and technical knowledge. The resource will depend on the number of audits scheduled and their scope and frequency. Auditors should have technical knowledge of the food products and processes being audited. Auditors cannot objectively audit their own work, so internal audits should be carried out by competent auditors who are independent of the area or activity being audited. This will ensure the objectivity and impartiality of the audit process.

7.8   The audit programme should be defined in an audit schedule that defines audit criteria, scope, the proposed auditor and planned frequency of audits. A procedure should be developed defining the requirements for the internal audit programme including planning and conducting audits, mechanisms for reporting the results of the audit and how any required actions will be monitored and followed up. Individual responsibilities should be defined within the procedure. The procedure should also outline the working documents that will be used such as audit reports, audit checklists and corrective action plans.

7.9   The results of the audit should be documented and brought to the attention of the management responsible for the area being audited. This should include both areas of conformance and non-conformance. Any corrective actions and timescales for their implementation should be mutually agreed. The management responsible for the area being audited is responsible for ensuring that any required actions are undertaken in a reasonable time frame in order to eliminate non-conformance and ensure effective corrective action. Verification activities should then be undertaken to ensure that the prescribed corrective action has been implemented and has been effective. A process should be put in place to manage all corrective action required within the organisation. This can be through a corrective action plan, non-conformance or corrective action log or similar document. The quality control manager should be responsible for monitoring the completion of corrective action according to agreed timescales and should highlight poor performance for follow up, additional resource and/or further action.

7.10   The implementation of quality assurance processes within an effective QMS/FSMS requires not only the implementation of procedural audits but also those that address extrinsic food safety hazards. This requires an audit plan to be established to monitor the 'fabric' of the manufacturing premises in terms of the building, equipment and tools premises; housekeeping and hygiene; and personnel hygiene. The frequency of these audits should be based on risk assessment including assessing the degree to which the product is enclosed within food processing equipment. These premises and personnel audits should be undertaken by trained, competent individuals and where possible should not just be seen as the completion of a checklist and an opportunity to identify non-conformance, that is, a tick-box activity. Auditors should be encouraged not only to have awareness of the premises and personnel standards required but also to use the internal audit process to improve staff understanding of the good manufacturing practice (GMP) standards required and promote continuous improvement. Senior management should support this activity as a preventive approach to minimising food safety risk.

7.11   Trend analysis should be undertaken across a series of internal audits to identify areas that give rise to ongoing, albeit minor non-conformance especially where corrective action is shown to have limited value in addressing weaknesses or breakdowns in the QMS and/or FSMS. This should form an input into the management review process as previously described. The trend analysis should form a framework for the development of specific key performance indicators (KPIs) or critical success factors (CSFs), which are robust enough to drive continuous improvement.

7.12   Guidance on undertaking audits and developing auditing programmes can be accessed in BS EN ISO 19011:2011 Guidelines for auditing management systems. Although not relevant to internal auditing itself ISO 22003:2007 *Food safety management systems—Requirements for bodies providing audit and certification of food safety management systems* provides guidance relevant to the undertaking of audit activities that still proves useful in developing an internal auditing programme. The annexes of this standard provide specific guidance on the time requirements for auditing.

8   PRODUCT AND PROCESS DEVELOPMENT AND
        VALIDATION

Principle

Products should be so designed as to ensure that the end product meets consumer expectation within the intended and anticipated circumstances of use, and to ensure that product design and performance have been fully evaluated for the required function in respect of microbiological safety, chemical safety, physical safety and sensory quality as elaborated in Chapter 2. To this end, a hazard analysis critical control point (HACCP) study undertaken by a multidisciplinary team is recommended, preferably applied from the earliest stages of product or process development with a view to eliminating or minimising potential hazards wherever possible and incorporating effective control parameters into the product design (see Chapter 3). The HACCP study should encompass any possible deviations from the design objectives. It is important to consider all realistic potential hazards, which may occur at each stage of the manufacturing process. In order to design the product to achieve these objectives, it is necessary to identify all the possible causes of the identified deviations. Hazard analysis and operability study (HAZOP) is a recognised hazard analysis method used during process development and validation to identify both food safety hazards and potential operational issues, which could lead to food safety or environmental hazards or impact on manufacturing efficiency (see Chapter 3). Similar considerations apply where changes are made to existing products, which would affect integrity, safety or stability of a product. Such changes could include those made to:

(a) ingredients or supplier;
(b) formulation or recipe;
(c) operations, for example, order of addition;
(d) machinery;
(e) processes or process parameters;
(f) packaging;
(g) storage;
(h) distribution;
(i) organisational structure or staff responsibilities; or
(j) consumer use.

The above list is not exhaustive but is based on experience of situations where failure to take such changes into account has resulted in serious human or commercial consequences or both.

Customer Requirements

8.1   Before commencing the design procedure, it is advisable to define the customer and/or consumer expectations of the new product or process. These should be established by market research or by direct

Food & Drink – Good Manufacturing Practice: A Guide to its Responsible Management, Sixth Edition.
The Institute of Food Science & Technology Trust Fund.
© 2013 John Wiley & Sons, Ltd. Published 2013 by John Wiley & Sons, Ltd.

agreement with the customer (e.g. the retailer, caterer or distributor). If established by market research, these expectations should be translated clearly into design objectives. It is also important to consider whether the food is targeted at a section of the population, which is deemed as a vulnerable group with regard to food safety, for example, the elderly, the young, pregnant women or immunocompromised individuals and the possible implications in terms of product design. Customer requirements may include the utilisation of recognised food safety hazards, for example, allergens in the product formulation; a formulation, packaging design or manufacturing process that may lead to higher risk of pathogen survival; or the use of packaging that in itself presents a food safety hazard, for example, glass, brittle plastic or metal.

8.2   Due account should be taken of any specific standards that are to be met to justify special claims (including those related to packaging), specified requirements such as suitability for a specific nutritional or health purpose, or other particular expectations, for example:

- environmental 'friendliness' recycling, genetically modified organism (GMO);
- free from specific ingredients, for example, sugar free;
- specific production method, halal, kosher, organic, vegan, vegetarian;
- animal welfare friendly;
- diabetic product; or
- low carbohydrate, low fat or suitable for a low-calorie diet.

These should be agreed before beginning the design process.

Design Objectives   8.3   Before commencing the design of a new product, design objectives should be established clearly, having due regard to the following requirements:

(i)   customer/consumer expectations;
(ii)   product safety;
(iii)   regulatory requirements for the intended market; and
(iv)   minimising operating costs while remaining consistent with the fulfilment of (i), (ii) and (iii).

Design objectives should include acceptance criteria and tolerances. Establishing tolerances is crucial to achieving customer expectations at a realistic cost. They may be established by past precedent or by agreement with the customer (e.g. the retailer, caterer or distributor).

Regulatory   8.4   Where regulatory requirements include statutory maxima or minima
Requirements   for particular parameters, the design targets should provide at least sufficient margin below the maxima or above the minima to allow for ingredient composition variation, within and between batch variation, and process variation.

Planning	8.5	Planning for a new process or product should include clear identification of responsibilities for the design activities, agreed actions and timescales. This is best achieved through a multidisciplinary project team, which could comprise technical, production, engineering, marketing, and purchasing and distribution expertise. In the event of changes to existing products, packaging or processing methods, the existing HACCP plan(s) and food safety management systems (FSMS) needs to be reviewed by the HACCP team and/or project team if they are the same group of individuals. Any proposed changes should be validated, and this should be done by a range of trials (see 8.6). When a new product is proposed, the HACCP team should consider whether the product is addressed by an existing HACCP plan, or whether a new HACCP plan needs to be initiated. In the event of a new packaging or process system being introduced, a new HACCP plan needs to be developed. These new HACCP plans, or a modified HACCP plan, must be approved by the HACCP team before factory trials can take place.
Quality Plan	8.6	At each critical stage in the design process [e.g. kitchen, pilot scale, pre-production factory trials, first production, packaging trial, distribution (transit) trial, retail trial], it is essential to verify that the product meets its agreed design criteria and agreed tolerances, and that the HACCP study has been successful to effectively manage food safety hazards. This can be achieved in a number of ways including:

- shelf-life testing;
- quality control testing of microbiological, chemical, sensory parameters and packaging;
- market trials;
- process monitoring; or
- checks for compliance with regulatory requirements.

The trials should be undertaken in accordance with documented procedures and the results of the trials should be documented and retained. If the factory trials and associated testing form part of the validation process for the HACCP plan and general FSMS, these should be referenced within the HACCP documentation. The Food Standards Agency (FSA) publication *E. coli* O157—Control of cross-contamination: Guidance for food business operators and enforcement authorities (2011) stresses the importance of not only validating the HACCP plan, but also focusing on validating the control measures in place to ensure that bacterial loading on fresh produce is reduced on receipt, that physical separation of materials is effective and that disinfectants are purchased and used in compliance with validated dilution levels and contact times. Traceability of materials used in factory trials is critical, and trial ingredients, in-process material and finished product should be suitably labelled, segregated and controlled to ensure that it does not enter the general stock system. If the materials used are, or contain, potential allergens, then allergen control procedures should be followed (see Chapter 4).

8.7 Nutrition analysis should be undertaken to ensure that at the end of the shelf life, nutrient values comply with the product label especially where the original design phase involved calculation of nutrition content from standard sources. This is critical where nutritional claims are made for the product, for example, free from, high in and low in.

8.8 A quality plan needs to be developed to manage key product quality parameters. Failure to achieve the design objectives can be caused by three factors: unrealistic initial objectives, an inadequate HACCP study or quality plan, or failure to implement the requirements properly. In all cases the entire procedure should be revisited, although revision of the initial objectives and design criteria may be considered if the end result is demonstrably acceptable in all respects despite the failure to meet the original target. If the objectives are changed at any point during the design and development process, a thorough re-evaluation of the implications of the changes must be made.

8.9 Guidance on the development of quality plans can be accessed in ISO 10005:2005 Quality management systems—Guidelines for quality plans.

Labelling 8.10 Checks should be undertaken to ensure that the labelling of the new product is legal and in accordance with the formulation and product specification. These checks should include weight control and volume control information, nutrition information, allergen information, customer storage instructions, cooking and preparation methods and other instructions for use, and product durability information.

9   DOCUMENTATION

Principle

Effective documentation is an essential and integral part of good manufacturing practice (GMP) and, in particular, one of the essential features of a properly operated hazard analysis critical control point (HACCP) plan and food safety management system (FSMS). Its purposes are to define the materials, operations, activities, control measures and products; to record and communicate information needed before, during or after manufacture; to reduce the risk of error arising from oral communication; and to form a vital part of the audit trail necessary for tracing components used to produce the final product. The system of documentation should be such that as far as is reasonably practicable the history of each batch of product, including utilisation and disposal of raw materials and packaging, intermediates and bulk or finished products, may be ascertained.

General

9.1   Documents fall into three main categories:

(a) manuals, which define company policy and protocols and how the organisation may seek to address a quality management system (QMS) standard such as EN ISO 9001:2000, or the British Retail Consortium (BRC) Global Standard for Food Safety;

(b) procedures, which define the company prerequisite programmes (PRPs) and procedures such as customer complaint procedures, management of incidents and product recall, traceability or calibration procedures; and

(c) work instructions, standard operating procedures, specifications and master forms. These are the documents that define production activities and identify the quality standards and the control checks that need to be undertaken. A master form is the blank template, which once completed becomes a quality record. These records can often form part of the organisation's 'due diligence' records.

Failure to undertake the required activities and complete the appropriate records will nullify many of the benefits of GMP and food control. Documents may include the following types:

(a) Manuals and policies including:
- quality manual;
- HACCP manual/plan;
- health and safety manual;
- Control of Substances Hazardous to Health (COSHH) manual (UK COSHH or equivalent elsewhere);
- quality policy;
- food safety policy;
- allergen policy;
- environmental policy;

Food & Drink – Good Manufacturing Practice: A Guide to its Responsible Management, Sixth Edition.
The Institute of Food Science & Technology Trust Fund.
© 2013 John Wiley & Sons, Ltd. Published 2013 by John Wiley & Sons, Ltd.

- ethical trading policy; and
- health and safety policy

(b) Procedures and PRP including:
- production programmes;
- training programmes;
- internal quality audit programme;
- pest control programme;
- plant maintenance programme;
- procurement procedure;
- supplier approval and performance monitoring programme;
- management of incidents and product withdrawal and recall procedure;
- quality control (including analytical and microbiological) procedures and methods;
- premises housekeeping and hygiene including cleaning schedules;
- traceability procedure;
- calibration procedure;
- personnel hygiene procedure;
- document control procedure; and
- quality records management procedure including archiving.

(c) Work instructions, standard operating procedures, specifications and master forms such as:
- ingredient specifications;
- packaging materials specifications including artwork standards;
- copy of order and/or terms of conditions of purchase;
- master manufacturing instructions, job notes and material resource planning instructions (including flowsheets and standard recipes);
- work instructions and task procedures;
- product formulations;
- intermediates specifications;
- bulk products specifications;
- finished products specifications; and
- machinery operating instructions.

(d) Records and reports (quality records)
These documents in particular should be designed with due regard to intelligibility to the intended user. This includes the need to recognise possible numeracy, literacy and language problems:
- records of monitoring of critical control points (CCPs) and quality control points (QCPs);
- records of receipt, examination, approval and issue for use of raw materials and packaging materials;
- records of the testing and release of intermediates, bulk products and finished products;
- records of process control tests;
- in-process recording instrument charts;
- weight or volume control charts;
- batch manufacturing records;
- customer complaint and service level reports;
- minutes of management system review meetings;

- quality control summaries and surveys;
- quality assurance, audit reports and records; and
- superseded documents.

(e) Additional QMS documentation
- site plans, for example, water system plans, plant layout, drainage plans;
- copies of applicable codes of practice, and relevant legislation; and
- equipment manuals.

9.2 Detailed advice on documentation related to specific production categories and specialised topics are contained in many of the chapters of this Guide. These chapters should be consulted in conjunction with this chapter.

Quality Manual 9.3 A quality manual should outline the structure of the QMS and provide a means to reference associated procedures, work practices, policies and forms. The quality manual must be established and implemented, and verification activities such as internal audits (see 7.7) must be undertaken to ensure its appropriateness to the processes undertaken and the degree of effectiveness in addressing food safety and quality criteria. The quality manual should be available to all relevant staff. It should integrate fully with the HACCP documentation and define how the organisational structure has been developed to ensure food safety and product quality and compliance with defined product and process criteria.

A quality policy is defined in EN ISO 9000:2000 as a statement of the overall intentions and direction of an organisation related to quality as formally expressed by top management. The policy defines the organisation's intentions and commitment to manufacturing safe and legal products that meet customer's quality expectations. The policy should be signed by the senior management individual with overall responsibility for the safety, legality and quality of the products manufactured. The organisation must demonstrate that it has communicated its quality policy to all personnel and that they understand the importance of the policy statement and associated QMS.

A food safety policy addresses the food safety intentions and objectives of the business and can be a standalone document or be integrated with the quality policy into a joint statement. Equally the quality manual may be a standalone manual that interfaces with the HACCP plan or a combined food safety and quality manual.

The objectives should be formalised, that is, documented, time based and measurable so that achievement can be demonstrated. The objectives, whether through the formal policy or other means, should be communicated to all staff. Consideration should be given to reading or language difficulties of staff when these objectives are communicated. Progress towards the achievement of food safety and quality objectives should be monitored by appropriate personnel at a minimum time frame of annually and the results form an input into the management review process.

Delivery/Intake Inspection Records   9.4   Due to the requirements of the supplier and buyer, and the need for inspection forms containing certain information along with other recorded data, a separate form is normally required for the inspection of every delivery or part thereof. Copies of this form may go to the vendor, production manager, purchasing manager and quality control manager. This documentation will also pay a pivotal role in the case of contract manufacturing (see Chapter 24). Considering the number of times this form is used and completed, it is necessary to make it convenient and brief, but complete in all the essential details relating to the consignment. This form will also be an important link in effectively managing product identification and traceability (see Chapter 10). Consideration should be given to designing the form in such a way as to remove the need to continually repeat certain information on every entry, for example, date, location and units of measurement, product that may in some circumstances only need to be stated once on the form. Computerised systems are being increasingly used to aid the speed and flow of information and allow for more rapid and convenient trend analysis. Traceability of raw materials and packaging materials to their initial inspection records is critical within the QMS/FSMS to ensure that food safety and quality criteria have been met. These records will form part of the audit trail in the event of a product withdrawal or recall (see Chapters 10 and 19).

In-Process/ Production and Finished Product Inspection Records   9.5   It is vital that accurate records be maintained of all relevant production data including actual processing details and performance, and end product quality. These records should be readily available for a predetermined time after the end of the expected shelf life of the product (see Control of Records Procedure, Section 9.9).

The manner in which these records are kept will depend on the requirements and size of the operation and can range from individual worksheets and diaries to full-scale computer systems. Any mandatory requirements, such as those for the UK Weights and Measures legislation, must be implemented (see 9.9). Traceability of in-process and finished products to the appropriate inspection records is critical within the QMS/FSMS to ensure that food safety and quality criteria have been met. These records will form part of the audit trail in the event of a product withdrawal or recall (see Chapters 10 and 19).

9.6   The use of clear concise notations should be ensured and all records made in a non-fade permanent ink. The use of Tippex-like materials should not be permitted; any alteration of an entry should be done by strike-through, accompanied by the corrected entry and initials of the person making the correction. Any amendments that require formal authorisation, for example, by the quality control manager, should be appropriately and clearly authorised. Wherever possible, all documentation should declare the units of measurement and the physical conditions of these measurements. The use of the British

Standards Institution (BSI) notation is recommended. Documents should, wherever possible, bear details of the target and actual values recorded.

9.7   Personnel completing records should have sufficient training on how to complete the form, and the effectiveness of this training should be assessed routinely. Particular attention should be paid to ensuring that personnel understand the need to fully complete the record. Personnel must be aware that a blank 'box' on the record implies that the inspection or test has not been undertaken, that is, there is no evidence to support it has been undertaken. The quality control manager should ensure that all quality control personnel show consistency in how they complete the records and also assess the use of quantitative rather than semi-quantitative or qualitative measures of quality criteria. The use of scoring systems or colour-coding systems (such as a 'traffic-light system') for records should be assessed, especially where criteria are being used to provide data to monitor performance against food safety and quality objectives. Terms such as 'good', 'nice' and 'ok' are qualitative and may represent in practice different standards of quality as assessed by individual staff. Appropriate training associated with developing quantitative measures to assess product quality, should drive greater consistency between individuals, shifts and processing units.

Training of production and quality control personnel should emphasise the importance of completing production and quality control records at the time the activity is undertaken for which they are recording the data. Personnel must not fill in a succession of ticks and/or initials at some later convenient time. Records should be designed where possible for actual numbers, code letters or words to be written rather than just a tick or cross. Repetitive ticking of a form can lead to checklist fatigue for the personnel completing the forms, and this must be avoided at all costs. It is very important that the individual initials each set of information that (s)he completes on a record so that in the event of non-compliance, there is traceability on the records to the particular individual who undertook the task. This will also provide an audit trail to training records in order to confirm the training that individual has received and whether it has been appropriate and remains current. Personnel, such as the quality control manager or designate, who are required to countersign or 'sign off' quality records should formally identify whether the countersignature identifies the form has been received and is ready for filing or if the countersignature demonstrates that the information on the record has been independently verified and signed off as complete.

Master Forms   9.8   To assist in the design and construction of record forms along with related documents, the following checklist may be of assistance for the quality control manager:

(a) determine person(s) responsible for developing and setting up the necessary forms and determine the competent personnel who will be undertaking the checks and then completing the records. Suitable training must be given to ensure adequate understanding of the role required of inspection personnel and the significance of identifying system or product non-conformance and the action to be taken;

(b) ensure that results may be recorded easily and there is sufficient space on the document to adequately record the results of inspections;

(c) ensure that results and data may be transmitted promptly;

(d) arrange that results and data be transmitted to the proper and appropriate personnel and that the recipients are in a position to take action as and when necessary;

(e) ensure there is sufficient space to record preventive and corrective actions taken as a result of inspection;

(f) ensure that re-inspection activities, if required, can be appropriately recorded;

(g) control charts may frequently be used to replace tabulated forms with the major advantage that the situation may be noted at a glance and not lost within a mass of numerical values. Where appropriate, data logging for automatic control systems should be used (see Chapter 28);

(h) wherever possible case markings, batch numbers, delivery vehicle and container numbers, along with container seal numbers, are all items of information that should be recorded when examining incoming and outgoing goods. All documents, samples and related data should be clearly dated, if necessary with the time, day, month, week number and year. The use of a sequential or consecutive numbering system is to be commended;

(i) ensure that if samples of packaging and labels are going to be attached to the forms, there is an indication of where they should be attached, the time they were attached, that is, whether quality assurance or quality control processes are adopted, and a mechanism for ensuring that such labels are formally signed off; and

(j) ensure that forms are routinely reviewed to ensure that the design of the form does not impinge on the ease of completion.

Control of Records Procedure   9.9   Not all records may need to be kept after the production period has ended. The QMS/FSMS should not require the retention of excessive records and should ensure that a protocol is set up and followed as to how long individual records should be kept, that is, for a defined time period. Special attention should be given to the shelf life of the product and any legal requirements, including provision of evidence of due diligence. A controlled records list involving an ongoing and continuously monitored system for purging the files of unwanted old data should be adopted. Control of records procedures should identify how records pertaining to the QMS/FSMS and 'one-up' 'one-down' traceability are collated, reviewed and verified, and maintained including storage and retrieval. In the event of a product recall, customer complaint or other incident, it may be required to retrieve records in a short period of time, for example, less than 2 hours, less than 4 hours or within a day, and this should be stated within the procedure and tested a minimum of annually as part of the product recall testing process. All records and data should be carefully examined by a responsible person before filing and at the same time screened for any irrelevant data, which should be removed. The retention time for documents and records should be determined by assessing the shelf life of the product or the shelf life of the customers' product if manufacturing ingredients or the product is suitable for freezing by the consumer. Any alterations to records should be authorised and dated, and justification for the alteration should be recorded (see 9.6).

Document Approval, Amendment and Control Procedure

9.10   A working method should be established and documented to ensure a document approval procedure is in place prior to issue, and only those personnel with the appropriate level of authority have approved the content and authorised the release of the document.

9.11   A further working method should be established and documented to ensure that all documents, specifications and so on, are updated, that any current amendments are maintained, and superseded documents are removed from the system. The reason for any changes or amendments should be recorded so that there is a suitable audit trail in place.

9.12   An adequate indexing and recall system, including records of amendments, is essential. The use of systems such as master indexes or master logs is recommended as a means to identify the current version of individual documents.

9.13   The working method should also include the method of document identification and the system for replacing existing documents when they have been superseded especially where there are multiple copies in circulation.

9.14   Documents should remain legible. Care should be taken when photocopying master copies to ensure that neither the quality of print is lost over time, nor a portion of the document is 'lost' due to incomplete copying; for example, the document control information is not copied that is on the bottom of the document. Attention should be paid to the number of copies of a document required so that the numbers reproduced will be used in a timely fashion and will be adequately controlled. This will minimise the potential for obsolete documents being used to record information.

9.15   Internal audits should be routinely undertaken to ensure that the appropriate version of documents is being used.

9.16   Documents received only by fax (other than on plain paper) are liable to fade and should be copied into permanent form if a long-term record is required. Documents received and retained in electronic format should be adequately controlled to ensure they are not inadvertently deleted or amended (see also 9.18).

9.17 Batch and laboratory samples when stored should be treated very much in the same way as documentation, records and data, but special emphasis must be placed on recording the physical conditions of storage, for example, temperature, humidity and light intensity. All labels should be firmly affixed to the container and be checked to ensure that they are durable for the period of expected life. The date on which they may be discarded should be clearly defined.

9.18 Provision should be made within the procedures to ensure that there is a back-up procedure in place for all electronic data to minimise the risk of losing data. Fire risk must also be assessed, and contingency plans need to be in place with regard to paper-only systems such as a fireproof safe for master copies/key documents (see Chapter 28).

10   PRODUCT IDENTIFICATION AND TRACEABILITY

Principle

It may not necessarily be enough to assume that the description of a consignment of a raw material on the packages or on the corresponding invoice is accurate. Where the identity is not absolutely obvious beyond question, the identity of each consignment of raw material should be checked to verify that it is what it purports to be. Product identification and traceability systems must be able to trace materials from source to finished product and also from product to source. The system should be routinely tested to demonstrate that it is implemented and effective. Product identification and traceability procedures are a key prerequisite to ensure good manufacturing practice (GMP).

Product Identification   10.1   Examples of raw materials, the identity of which may be obvious beyond question, are many fruits and vegetables (although where a particular variety or cultivar is specified, the manufacturer receiving a consignment will need to determine that it is of the cultivar specified).

10.2   Major examples illustrating the dangers of simply assuming that raw materials are what they are supposed to be, have been the inadvertent use, by some manufacturers, of what subsequently was revealed to be unfit raw materials that have subsequently contaminated food batches or where an animal feed manufacturing operation is involved has led to the slaughter of animals in the supply chain who have been potentially fed contaminated feed, for example, dioxin contamination of feed.

10.3   In evaluating a supplier's premises and operations, attention should be paid to, and account taken of, the safeguards operated by the supplier against mistaken identity of materials supplied.

10.4   In some instances, checking a raw material delivery for compliance with its specification, as indicated in 6.12, will also establish its identity. Where this is not the case, consideration should be given to building into the inspection schedule a check to verify material identity. It would be a mistake to assume that because an ingredient may be a clearly defined chemical entity, supplied by a reputable supplier who gives assurances as to compliance with legal purity requirements, that its identity may be taken for granted. Chemical manufacturers who manufacture food additives often also manufacture other chemical substances for non-food use, and mistakes or mislabelling could occur. Because of the seriousness of the potential hazard involved in such an occurrence, it is all the more important that each consignment of an additive should also be checked to verify that its identity and that it is 'food grade'. In the instance that fruits and vegetables are imported, then on receipt it is important to assess that the designated country of origin (COO) is recorded and it is verified that it is correct. Where imported food is processed in establishments approved through either legislative means or third-party certification, it is important that product identity is established on physical labelling and associated documentation and is verified on receipt.

10.5 In practical terms, for traceability to be effective, the responsibility of each organisation is to ensure the 'link' in the food supply chain is not broken. This requires an organisation to satisfy itself that its supplier of a food material has carried out checks to ensure that the documentation provided accurately describes the nature and substance of that material and its provenance (traceability/lot number). The organisation should retain that documentation and, following the manufacturing process, issue documentation/labelling to the organisation to which they in turn sell the food product.

Traceability
Procedure

10.6 A documented procedure should be developed and implemented to ensure that, throughout all stages of a company's operations, raw materials and in-process and finished products are identified and traceable. Packaging and processing aids should also, where practicable, be traceable to an individual delivery and/or batch. The quality control manager is responsible for the implementation and validation of this procedure. Guidance on the general principles and basic requirements for traceability system design and implementation is contained in ISO 22005:2007.

10.7 Where reworking/repacking operations are being performed, traceability must be maintained through quality control documentation.

10.8 The quality control manager or designate should ensure that the product identification and traceability procedure should be tested at prescribed intervals. These tests should interface with the product recall and withdrawal procedure tests and should include a variety of potential scenarios (see 19.9–19.11). This test should include a mass balance or quantity test, that is, that the volume/weight of the product in the designated batch can be fully traced from intake to despatch or vice versa. This requires any reworked, regraded or waste product to be fully recorded within manufacturing records. Each manufacturing unit should undertake a risk assessment to determine the level of traceability achieved in practice and whether in the event of a product recall further batches should be automatically recalled. This risk assessment should also include a determination of the timescales that are acceptable for the traceability audit, for example, what documents should be available within 2 hours, 4 hours, 8 hours and so on (see Chapters 9 and 19).

Lot Marking
Regulations

10.9 The **Food (Lot Marking) Regulations 1996** set out the lot marking requirements to be applied to all foodstuffs sold for consumption (unless specifically exempted) including wines and spirits. They implement **Council Directive 89/396/EEC** on indications and marks identifying the lot to which a foodstuff belongs as amended by **91/238/EEC** and **92/11/EEC**. The directive establishes a framework for a common lot (or batch) identification system throughout the community in order to facilitate product recall along the whole of the food chain (see 19.8).

Genetically Modified Organism (GMO) Regulations   10.10   **EC/1830/2003** requires that all genetically modified (GM) ingredients are identified and labelled at their point of production and that this information remains present throughout all stages of the food chain. **EC/1829/2003** includes requirements for the labelling of food and food ingredients consisting of GMOs or produced from GMOs, such as sugars, oils and starches.

Provenance, Identity Claims and Assurance Status   10.11   Where claims are made on the finished product, it is important that the ingredients or the product itself comes from a designated provenance, or has a designated identity or assurance status, for example, Fairtrade, organic, COO. The organisation must have a suitable identification and traceability procedure in place to demonstrate the status of raw materials, in-process material and finished product. Records should be maintained to demonstrate that the finished product complies with the legal requirements associated with such claims.

11   PERSONNEL AND TRAINING

Principle

There should be sufficient personnel at all levels with the ability, training, experience and, where necessary, the professional and technical qualifications appropriate to the tasks assigned to them and commensurate with the size and type of the food business. Their duties and responsibilities should be clearly explained and recorded as job descriptions, with associated organisation charts or by other suitable means. Training should cover not only key tasks, but also good manufacturing practice (GMP) generally, and the importance of, and factors involved in, personal hygiene. Personal hygiene protocols are a key prerequisite to ensure GMP.

General

11.1   Manufacturing companies vary widely in character, size and structure, and whether they are independent or part of a group, and, in the latter case, the extent to which the character of their operations is subject to central control. There are also wide differences in terminology and job titles in different companies. These circumstances make it impossible to adopt rigid generalisations about the titles of key personnel but the following principles are of importance, and the terms 'production manager', 'quality control manager' and 'purchasing manager' are used here for the sake of convenience, but may equally well be read as 'general manager', 'operations manager', 'technical manager', 'quality assurance manager', 'procurement manager' or as the departments concerned.

11.2   Two key personnel are the production manager and the quality control manager, who should be two different persons, neither of whom is responsible to the other, but who both have a responsibility to collaborate in achieving product safety, legality and the required product quality. A third key person is the purchasing manager, who has a special responsibility for ensuring that raw materials/ingredients are purchased in compliance with specifications, and for collaborating to that end with the quality control manager. A supplier quality assurance protocol that includes supplier performance monitoring should be developed for the manufacturing company following collaboration between the purchasing manager and the quality control manager (see Chapter 16).

11.3   Persons in responsible positions should have sufficient authority to discharge their responsibilities that should be clearly defined (see Section 11.8). In particular, the quality control manager should be able to carry out his/her functions impartially and in accordance with the Institute of Food Science & Technology (IFST) Code of Professional Conduct (http://www.ifst.org).

Food & Drink – Good Manufacturing Practice: A Guide to its Responsible Management, Sixth Edition.
The Institute of Food Science & Technology Trust Fund.
© 2013 John Wiley & Sons, Ltd. Published 2013 by John Wiley & Sons, Ltd.

11.4 The quality control manager should possess appropriate professional and scientific/technological qualifications, and experience in food quality control. (S)/he should also possess interpersonal skills so that (s)/he can effectively communicate food safety and quality criteria at all levels within the organisation.

11.5 Key personnel should be provided with an adequate number and calibre of supporting staff. A formal human resource appraisal should be undertaken with consideration given to the requirements of developing, validating, implementing, monitoring, revalidating and verifying the hazard analysis critical control point (HACCP) plan(s), the food safety management system (FSMS) and quality plans. The resource appraisal review should also include identification of the human resource required to implement effective systems. A review should be undertaken minimally each year, or when operational circumstances change, which impact on the resource requirement.

11.6 Key personnel should be direct employees of the company. Persons contracted, but not employed directly by the company, may be retained to provide advice or to carry out special projects, but should not be regarded as filling key personnel positions. During the approval process for contract personnel, it is important to determine the level of qualifications, current validity of membership of professional bodies and the status of continuous professional development (CPD), and wherever possible, seek references from previous clients.

11.7 Persons should be designated to take up the duties of key personnel during the absence of the latter. This should be formally documented, and the organisation must be able to demonstrate that deputies are competent to undertake the role when required.

11.8 The ways in which those responsibilities that can influence product safety, legality and quality and the authority to discharge them are distributed among key personnel may vary with different manufacturers. What is essential in all organisational arrangements is that responsibilities should be clearly defined and allocated and areas of authority delineated. These arrangements should be properly understood and respected by all concerned. This may be documented in job descriptions/specifications or organisation charts or alternative documents.

11.9 The quality control manager should have the authority to establish, validate, implement and verify all food safety and quality control procedures; to approve materials and products; and to withhold approval from any material or product that does not comply with the relevant specification. In some organisations, withholding of approval amounts to absolute rejection; in others, it amounts to provisional rejection subject to confirmation by the general management or the managing director; in yet others, it may be absolute rejection in certain defined serious cases, provisional rejection in others. In exercising authority in the foregoing ways, the quality control manager, subject to the proper discharge of responsibilities, is under obligation to do everything possible to minimise and preferably prevent any disruption of production and/or distribution arrangements and schedules; and to provide prompt information and advice to production personnel to help maximise conformance with specification, and, in the case of non-conformance, to provide advice on rectification or reworking if possible. The quality control manager should, where required, identify whether the rejection is on the grounds of legislative, food safety or quality criteria to ensure suitable advice to production and appropriate corrective action. The quality control manager has a leading role in defining preventive action where weaknesses in the quality management system (QMS) and/or compliance in practice with the QMS have been identified and the non-conforming product has not yet been produced. The production manager should fully support such initiatives and enable preventive action to be implemented and be effective.

11.10   The production manager should have responsibility for manufacturing resources including processing areas, personnel, equipment, operations and records, and for using these resources to produce, in accordance with the master manufacturing instructions or equivalent, the programmed quantities of products conforming to the relevant quality specifications and within the budgeted cost. In addition, the production manager usually has some responsibilities that are shared, or exercised in liaison with, the quality control manager and other personnel. The production manager should be a member of the HACCP team and should participate in the development and acceptance of specifications, particularly the master manufacturing instructions and the finished product specifications, and in the training of production personnel; (s)/he should liaise with the quality control manager in ensuring an optimum manufacturing environment, in particular in relation to controlling and minimising extrinsic hazards (food safety hazards arising from the processing area, personnel, equipment and operations), hygiene and for quarantining and rectification of any non-conforming product. The production manager should not have authority to use any raw materials or intermediates that have not been approved for use, unless a concession has been formally agreed with the quality control manager or designate.

Supplier Quality Assurance

11.11   The quality control manager should have an advisory role in the choice of suppliers of raw materials, packaging and services (see Chapter 16).

Recruitment and Selection

11.12   In the recruitment and selection of staff, due regard should be given to the required prerequisite skills and the potential suitability of the candidate for the task in hand. That is to say, for example, persons of clean and tidy and methodical nature may be more likely to appreciate the principles required in the hygienic handling and storage of food and the implementation of the QMS, than those who are not. Similarly potential employees should be appropriately qualified for the requirements of the task as defined in the job description/specification.

11.13   Training is necessary to enable personnel at all levels within the organisation to understand their responsibilities and improve their knowledge and skills. Effective training will also improve individual performance and reduce the level of supervision required. Training will only be effective when the provision of knowledge develops thorough understanding and is relevant to the individual's tasks and responsibilities. The implementation of the knowledge acquired by individuals requires management support including appropriate opportunity to practise skills, ensure motivation and provide effective supervision. It should be noted that while a certificate of attendance at an internal/external training course is of value, that value is limited if it cannot be verified that the person(s) attending such training can effectively implement the skills and knowledge gained in practice. Similarly signatures on a training form demonstrate training has taken place but do not in themselves provide evidence of ongoing competence. Therefore the internal audit programme needs to include in its scope the verification of personnel competence at all levels of the organisation where their work activities can affect food safety, legality and quality.

Training records should be maintained and they should provide details of the nature of the training undertaken, the date and duration of the training activity, the identity of the trainer and the signatures of the trainee to verify training has taken place and the trainer to confirm that the training was effective and the individual is competent. Individuals should only be signed off when they are deemed as competent and have demonstrated this in all aspects of the job role to the trainer. Some job roles will require specific training, for example, individuals involved in chemical handling, machine operating and pest control, and this must be documented and signed off by the relevant person that the individual has attained a minimum level of competence. The initial and ongoing level of competence of the trainers should also be identified and documented by the manufacturing organisation in order to verify that they can provide appropriate training and knowledge transfer to staff. If agency staff are used, then their training and the competence of the agency trainers must be verified as suitable before they commence work. Additional site-specific training will be required in addition to familiarise agency staff with the specific aspects of the manufacturing process and the food safety, legislative and QMSs in place on the site. Full consideration should be given to the use of agency staff at critical control points (CCPs), critical quality points (CQPs) and critical legislative points in the manufacturing process as to their suitability and degree of competence for the role.

11.14   All production and quality control personnel should be trained in the principles of GMP, and in the practice and underlying principles of the tasks assigned to them. Similarly, all other personnel (e.g. those concerned with maintenance or services or cleaning) whose duties take them into manufacturing areas or bear on manufacturing activities, should receive appropriate training. Records should be kept of the training of each individual. Where individuals are working at

CCPs as defined in the HACCP plan and/or have responsibility for the effective implementation of prerequisite programmes (PRPs), the manufacturing company must develop and implement training programmes to ensure that these individuals have appropriate levels of competence, skills and knowledge to support the effective adoption of the HACCP plan. Similarly, those individuals working at CQP and critical legislative points such as label printing, quantity control (weight, volume, count) and points of control of provenance, assured status and identity control should have appropriate training and supervision to ensure that legislative requirements are consistently met.

11.15   Training should be in accordance with programmes approved by the production manager and the quality control manager. Training should be given at recruitment (usually termed induction training) and repeated, augmented and revised as necessary. The degree of an individual's knowledge and understanding of the content and application in the workplace of the information contained in the induction training should be tested and verified before individuals can commence work. Retraining should be undertaken if required. Records of induction training should be retained for all personnel, and with returning seasonal workers, induction training should be undertaken each season with particular emphasis on changes in legislation, customer requirements and emerging food safety hazards. Both in training itself, and with regard to the need for personnel to be able to understand and follow written instructions and procedures, notices and so on, particular attention should be given to overcoming language, numeracy or literacy difficulties. Recommendations for the content of induction and supplementary hygiene training programmes are contained within the Industry Guides to Good Hygienic Practice.[1] These guides were initially developed in accordance with **Article 5 of the EC Directive on the Hygiene of Foodstuffs (93/43/EEC). Regulation (EC) No 852/2004** stated that the guides developed in accordance with the Directive on the Hygiene of Foodstuffs (93/43/EEC) would continue to apply after 1 January 2006 if they remained compatible with the objectives of the new legislation. Many older guides may not necessarily cover all of the revised requirements, and to date, they have not all been reissued as revised guides. Six guides have been subsequently published that address the following sectors: mail order food; wholesale distribution; flour milling; vending and dispensing bottled water; and retail. While flour milling would be the only sector defined as a manufacturing process, manufacturing organisations need to be aware of the requirements of subsequent sectors in the supply chain that they supply in order to ensure supply chain compliance.

11.16   Periodic assessments of the effectiveness of training programmes should be made, and checks should be carried out to confirm that designated procedures are being followed. Training programmes can be evaluated by a number of techniques including:

---

[1]These are available from The Stationery Office (http://www.tsoshop.co.uk).

- pre- and post-training observation and questioning to assess if knowledge has been gained, practices or standards have improved or attitudes have changed;
- post-training assessment to determine the individual's competence to produce safe food, which may involve written tests, observation or questioning, microbiological swabbing of hands and/or food contact surfaces or bacterial food sampling; and
- trend analysis of the frequency of customer complaints, quality incidents or level of waste or rework or the frequency of failures at CCPs and CQPs.

11.17 Training should be reinforced by adequate supervision, mentoring and support, regular performance reviews, refresher training at designated intervals, posters and notices, reviewing the results of routine microbiological monitoring and quality performance with personnel and recognising good practice. Refresher training is required where any element of the QMS or FSMS changes, especially the induction programme, so that all staff are consistently trained to the same level. The time period as to when a person is deemed competent and thus needs less supervision is very dependent on the individual and requires ongoing review by the relevant member of management. The organisation should develop an electronic training database, or a paper-based or electronic training matrix to control the training requirements for the organisation. The training matrix, or equivalent, should state the current level of training for all staff. It should identify for a specific job role what level of training and competence is required in terms of internal and external training and the current level of training of all staff in that job role. This will assist the organisation in undertaking a training needs analysis to identify those who require further training for the first time or in the form of refresher training. Therefore the training matrix should identify those staff currently in the induction phase and receiving training; those individuals who are signed off as competent, but may need refresher training; and those who are deemed able to train others. The training matrix should be reviewed at designated intervals and if as a result of review and/or audit activities inconsistencies in training and competence have been identified. The details of the review and subsequent preventive or corrective action should be recorded and records retained as a demonstration of due diligence. Formal employee appraisals should be undertaken to review in part, ongoing performance and areas requiring further development and training.

Cessation of Employment

11.18 Where employment is terminated for any reason, consideration should be given to any possible risks arising through disaffection or simple lack of continuing interest and commitment from the employees concerned, and appropriate precautions taken as necessary.

Food Hygiene and Personal Hygiene

11.19 Many important aspects of hygiene are also dealt with in Chapter 12 (Premises and Equipment) and in Chapter 6 (Manufacturing Activities). It is essential that all personnel are appropriately trained in relevant aspects of food hygiene. **Regulation (EC) No 852/2004 of 29 April 2004 on the hygiene of foodstuffs** came into automatic force in all Member States on 1 January 2006, and continued the requirement of earlier legislation that food business operators should ensure:

1. that food handlers are supervised and instructed and/or trained in food hygiene matters commensurate with their work activity;

2. that those responsible for the development and maintenance of the procedure referred to in Article 5(1) of this Regulation or for the operation of relevant guides have received adequate training in the application of the HACCP principles; and

3. compliance with any requirements of national law concerning training programmes for persons working in certain food sectors.

Sections 11.20–11.22 refer particularly to ensuring high standards of personal hygiene of all concerned with the production processes and all persons entering production areas, including quality control personnel, product development personnel, engineers and maintenance personnel, inspectors, senior management, contractors and visitors.

11.20   Instruction should not be regarded as an adequate substitute for training, which provides not only information on what should be done, but an understanding of why it is important. Training should embody the requirements of personal hygiene, the reasons why they are important and the fact that they are legal requirements. Adequate facilities and resources must be provided to enable personnel to comply fully with those requirements.

11.21   In the UK, the manufacturer must comply with the requirements of **EU Regulation (EC) No 852/2004** and **EU Regulation 853/2004**, which are also tabulated in Schedule 2, 'Specified Community Provisions', in the **Food Hygiene (England) Regulations 2006** and in the equivalent Scotland, Wales and Northern Ireland Regulations (or with any regulations that may at any future time supersede these regulations). The **EU Regulation (EC) No 852/2004** states that every person working in a food-handling area is to maintain a high degree of personal cleanliness and is to wear suitable, clean and, where necessary, protective clothing but does not specify the detail of personal hygiene requirements. However the following should be regarded as essential:

(a)   every person working in a food-handling area shall maintain a high degree of personal cleanliness and shall wear suitable, clean and, where appropriate, protective clothing. It should be noted that the term 'protective' designates protection of the food from contamination and should not be seen in the context of personnel protection and safety;

(b)   no person, known or suspected to be suffering from, or to be a carrier of, a disease likely to be transmitted through food or while afflicted, for example, with infected wounds, skin infections, sores or with diarrhoea, shall be permitted to work in any food-handling area in any capacity in which there is any likelihood of directly or indirectly contaminating food with pathogenic micro-organisms, and a person who suspects that there is any such likelihood on his/her part must report it to the proprietor of the food business. Further guidance is available from the Food Standards Agency (FSA) publication Food handlers: Fitness to work regulatory guidance and best practice advice for food business operators (2009)[2];

(c) an adequate number of hand-wash stations must be available, suitably located and designated for cleaning and effective drying of hands. All sanitary conveniences within food premises must be provided with adequate natural or mechanical ventilation. An adequate number of flush lavatories must be available and be connected to an effective drainage system. If toilets other than non-flush-type lavatories are used, then a documented risk assessment must be in place to state that the use of alternative designs does not present a contamination risk to food handlers. This risk assessment must be validated and routinely verified to demonstrate that the levels of personal hygiene required in a food manufacturing operation are consistently achieved. Lavatories must not lead directly into rooms in which food is handled; that is, there should be a minimum of two doors with an intervening ventilated space. Signs must be displayed in all toilet area in a range of languages as required to inform people that they must wash and sanitise their hands before leaving the area. Hand-wash stations with adequate resources must be provided within the toilet areas as well as on entry to production areas. There must be adequate training and ongoing supervision to ensure that protective clothing is removed and suitably protected from contamination prior to entry into toilet areas. If showers are provided, consideration should be given to the potential contamination of shower heads with *Legionella* bacteria. A procedure should be established, implemented and verified to control the potential for *Legionella*;

(d) hand-wash stations must be provided with hot and cold running water, materials for cleaning and, if required, sanitising of hands, and for hygienic drying. The provisions for washing food and washing equipment must be separate from the hand-washing facility. Taps should not be present. Instead the water flow should be knee, foot or magic eye operated. The turning off of the water must be automatic. Signage should be at the entrance to all production areas requesting that hand washing is undertaken before entry. Signage on best practice for hand-washing should also be in place including schematics for how to wash, clean and disinfect hands especially in manufacturing environments where high-risk foods are prepared. Minimum contact times for chemical products on hands should also be identified. Instruction on effective hand-washing techniques should be given during induction and refresher training given as required.

[2]http://www.food.gov.uk/multimedia/pdfs/publication/fitnesstoworkguide09v3.pdf.

Hand-washing procedures should be documented, and personnel should wash hands on return to the production area and before commencing work; after handling waste or undertaking cleaning tasks; after visiting the toilet; and after all break periods. Where required, hand sanitiser should be used. The FSA publication *E. coli* O157—Control of cross-contamination: Guidance for food business operators and enforcement authorities (2011)[3] paragraph 54 to 60 recommends that for extra protection against cross-contamination, a liquid hand wash that has disinfectant properties conforming to the European standards BS EN 1499:1997 is used. Consideration should be given to cultural barriers to the use of alcohol-based products and suitable alternatives sought where required. Routine hand swabbing could be used to determine the effectiveness of hand-washing techniques. The temperature of the water used for hand washing should be routinely checked and recorded. The Campden Guideline G62: Hand hygiene: Guidelines for best practice (2009)[4] recommends a water temperature of between 35 and 45 degrees Centigrade. Signage should be in place to identify areas designated for hand washing only. The provisions for washing food and also washing equipment must be distinct and signed as such. Practice should be routinely verified to ensure that the correct activities are being undertaken at each specific designated sink/washing station;

(e) the prohibition of the wearing of wrist watches and jewellery except for plain wedding rings or wedding wrist bands in food production and storage areas. The company should have a defined policy on the type of jewellery allowed to be worn for medical, ethnic or religious reasons and the controls in place to ensure personnel health and safety as well as minimising the risk of product contamination especially if wedding rings and plain sleeper earrings are worn. Piercings including rings and studs in exposed parts of the body such as ears, noses, tongues and eyebrows should not be permitted;

(f) adequate changing facilities for personnel must be provided, where necessary. The changing facility must be designed for effective cleaning. These must provide adequate space for the number of permanent, agency and temporary staff and be designed to separate outside and inside clothing and provide secure storage for personal belongings. There must be provision for designated storage for clean and dirty protective clothing to prevent cross-contamination. Staff should enter the production facility direct from the changing room and not go outside before entry. Visitors and contractors may use the same changing facility or a designated facility; again they must enter the production areas directly from that facility. Lockers must not be used to store food items. A separate provision must be made on-site for storage of food brought onto site. The cleaning of lockers should

---

[3]BS EN 1499:1997 Chemical disinfectants and antiseptics. Hygienic handwash. Test method and requirements (phase 2/step 2).
[4]http://www.campden.co.uk/publications/pubDetails.php?pubsID=4480.

be addressed in the cleaning schedule and their maintenance should be undertaken as required. Lockers should be so designed that dirt, rubbish and waste material cannot accumulate either on top or underneath the lockers. Internal protective clothing should not be worn outside and should be removed on leaving a food production area and be suitably stored. Designated separate lockers may need to be provided for both internal and external workwear as well as personal belongings that must not be taken into the production and storage areas. In high-care food manufacture, a protocol area is recommended, which requires formal separation of external clothes and internal workwear including the adoption of separate lockers as well as a barrier in the room to formally segregate high- and low-care areas;

(g) an adequate provision of safety footwear, and suitable protective clothing and the laundering thereof. There should be sufficient protective clothing per person to ensure that appropriate changing of clothing can be undertaken. The frequency of changing of protective clothing must be determined in light of food safety risk assessment. This could be either on-site or third-party contract laundering of clothing. The contractor supplying laundry services must be subject to the supplier approval and monitoring procedures (see Chapter 16). Designated internal footwear such as shoes or boots should be segregated from external footwear. In high-care environments, personnel should understand why disinfection of internal footwear is important and follow the designated barrier procedures at all times;

(h) laundry procedures must provide for the appropriate degree of cleaning of protective clothing. There should be adequate segregation during storage and after laundry of clean and dirty clothing and clean clothing should be protected from recontamination. The temperature of the wash/dry cycle should be determined as to whether the clothing needs to be clean or in the case of high care, clean and disinfected through the temperatures achieved in the washing/drying cycle. Temperatures should be monitored on a routine basis to ensure they meet minimum requirements. The results of monitoring should be recorded. High-risk and low-risk area clothing must be laundered separately. Consideration should be given, based on risk assessment as to whether home laundry is appropriate. If home laundry is permitted, specific procedures must be put in place and staff should not be allowed to travel to work in their protective clothing. If company-issued clothing/uniforms are used that are permitted to be worn to work, a risk assessment must be undertaken to determine what additional protective clothing is required to protect food and materials from contamination. This is especially important in low-risk businesses that allow home laundry and the wearing of work clothes/uniform from home to work, for example, in despatch areas. Suitable provision must be made for smoking and eating of food to prevent contamination. Verification activities to determine the effectiveness of the laundry procedures must be implemented and action taken to address non-conformance if required;

(i) the wearing of clean protective overclothing, including headgear and, where appropriate, neck covering, by food handlers and any persons visiting food rooms. Protective clothing must be designed with no external pockets and fastened with press studs rather than buttons as well as falling to below the knee (see 5.8). Specific cleaning and/or disposal procedures must be developed for items of protective clothing, for example, plastic disposable clothing, chain mail aprons and gloves, and liner and outer gloves and coats used in chilled despatch and freezer stores. The cleaning of high visibility clothing and hard hats where it could come into contact with the product must also be determined and procedures implemented. Hoods on underclothing must be contained within the protective clothing at all times as well as any fastenings on underclothing that could pose a food contamination risk if dislodged. Scarves should not be worn in chilled and frozen temperature controlled areas as there is a personal health and safety risk. Where disposable protective clothing is used, procedures for adequate control should be developed based on risk assessment;

(j) pre-employment medical checks, so that no person suffering from, or a carrier of, any of the specified kinds of infection is employed as a food handler; the active encouragement of personnel to report infections and skin lesions; and the encouragement of supervisory personnel to look out for signs and symptoms of such conditions and the development of return-to-work procedures following illness or holidays abroad. Medical screening requirements will vary according to the types of product being manufactured and whether they are high or low risk. Pre-employment medical checks can include stool sampling to verify that individuals are not carriers of harmful pathogens before they commence work or be a simple structured series of questions. Examples of such pre-employment questionnaires are included in the FSA Food Handlers: Fitness to Work—Regulatory Guidance and Best Practice Advice for Food Business Operators 2009, published in May 2009. The quality control manager, or designate, should verify that the person is fit to work as a food handler, both on pre-employment checks and in the case of illness or holidays abroad. The individual should be competent to check the questionnaires and as a minimum hold a current Level 2 food safety certificate.

(k) the prohibition of carrying loose items in the production areas especially mobile phones. Where deemed necessary, pen control policies, as with any other loose items, for example, keys, should be in place to prevent loss or contamination of food (see 5.8);

(l) that eating, drinking and smoking should only be undertaken in a designated area. Protective clothing should be removed before breaks periods. Smoking while wearing protective clothing should be prohibited. If canteen/restaurant facilities are provided either directly or via a catering supplier, there must be a HACCP plan in place that addresses the provision of food. In the UK the Safer Food Better Business (SFBB) Manual may have been adopted within the kitchen facility. The FSMS and the associated records should be reviewed on a routine basis by the quality control manager or designate. The personnel hygiene requirements identified in this chapter should apply to the canteen/catering staff as equally as food production staff. If canteen/restaurant facilities are not provided on-site, and staff bring in their own food, they should be provided with adequate designated refrigerated storage facilities that are clean and hygienic and capable of maintaining appropriate temperatures. Fridge air temperature should be checked on a routine basis to ensure compliance. If cooking facilities are provided, including microwaves, they as well as the fridge should be identified on a cleaning schedule and a cleaning programme developed, implemented and verified. Particular attention should also be given to the cleaning and disinfection of food contact surfaces that could be vehicles for food contamination.

(m) that perfume, scented deodorant or aftershave should not be worn as this could potentially taint the product. For this reason perfumed disinfectants or hand soaps should also not be used;

(n) that all hair should be fully contained to prevent product contamination. This includes head covering, hairnets and snoods as appropriate. Consideration should be given to the containment of body hair on arms, torso and so on, and to this end, protective clothing should come down to the wrist and fasten to the neck. The Campden BRI Guideline G48: Guidelines for preventing hair contamination of food (2006)[5] provides further information on the protocols that should be in place;

(o) that gloves should be controlled to prevent them from being a source of foreign body contamination. A glove issue and control procedure should be in place and awareness training undertaken so that personnel are instructed that the gloves are worn to protect the food from contamination not to keep their hands clean and that a gloved hand is as capable of causing cross-contamination as a bare hand. Latex gloves should not be used due to the risk of allergenic reaction by individuals allergic to latex. If a part of a glove, or indeed the glove itself, is lost, then the quality control manager, or designate, should be notified immediately so that appropriate corrective action can be implemented (see 11.21q for corrective action procedure in the event of the loss of a glove or plaster);

(p) that fingernails should be kept clean, short and unvarnished. False fingernails or nail varnish should not be worn as they could prove a source of foreign body contamination. False eye lashes and/or excessive make-up should not be worn. Consideration should be given, depending on the food products being manufactured, to formal checks being undertaken before work commences, to prevent contamination;

(q) all cuts and grazes on exposed skin should be covered with a company-issued detectable coloured metal strip plaster. They must be issued and signed out and checked at the end of produc-

[5]http://www.campden.co.uk/publications/pubDetails.php?pubsID=98.

tion to ensure they are still in place. The batches of plasters, where appropriate, should be regularly tested through a metal detector. Wounds obtained out of working hours (in particular, those to the hands, arms and face) should be reported to the relevant manager at the start of work. Normal plasters must then be removed and replaced with company-issued dressings. First aid kits must be under formal control and be monitored to ensure that all items stored within the first aid kit and when they are used can be fully accounted for. The stock in the first aid kits must be routinely verified to ensure effective control. It is accepted practice to determine at the end of the shift, or day's production as appropriate, that any plasters issued are accounted for. The loss of a plaster, or a piece of or a whole glove, must be reported immediately to the relevant manager and the following actions taken:

- production should be stopped;
- a search must be undertaken of the immediate location, general workplace and waste containers for the plaster as appropriate;
- if the plaster cannot be found, then packed product must be re-inspected;
- if the plaster is still not found and there is the potential for the item to be in a consignment that has already been despatched, the quality control manager or designate must be notified immediately and appropriate corrective action implemented.

(r) staff should inform managers or supervisors if they are required to take personal medicines especially for long-term conditions such as for allergies, diabetes or asthma. A personal medication procedure should be in place to control all medication brought onto site in order to prevent inadvertent contamination of the product and the procedure must be routinely audited to determine its effectiveness.

(s) visitors and contractors should undergo medical screening before being allowed entry to production areas. Where pre-entry medical questionnaires are completed by visitors and contractors, they should be formally reviewed and signed off by an appropriate member of staff before the individuals are allowed to enter a production facility. Visitors and contractors should be accompanied by staff, or otherwise controlled, at all times to ensure that personal hygiene requirements and product protection controls are fully complied with and security arrangements are fully met;

(t) verification activities should be undertaken on a prescribed time interval basis to ensure that personnel are fully complying with all the above requirements. Programmes of refresher training for employees should be undertaken at regular intervals, and as required, if issues of poor personal hygiene practices are identified.

Personal Hygiene 11.22 It is the responsibility of the quality control manager to ensure that
Policy personal hygiene policies and protocols have been developed and that

good hygienic practice has been explained to, and is fully understood by, employees, visitors and contractors. Managers should communicate the policies and protocols to all personnel and demonstrate full management commitment to the requirements; advise employees, visitors and contractors of their legal obligations under European Union (EU) and UK food hygiene legislation to report any infectious or potentially infectious conditions; and exclude infectious or potentially infectious food handlers as specifically required by EU legislation. Managers should continuously review whether there are any barriers to good personnel hygiene practice and remove them where necessary. These include ensuring that:

- the design of the premises and positioning of hand-wash stations allows for adequate hand washing and sanitisation if required;
- there are adequate supplies of clean protective clothing and hand-washing sundries at all times;
- staff have sufficient time available to undertake good hygienic practice;
- the personal hygiene training given, including the training aids used, is appropriate to the language skills and literacy of the staff. Where appropriate, pictorial training should be used; and
- staff, visitors and contractors have sufficient understanding of the potential effects of poor personal hygiene.

Managers should follow appropriate action including retraining, with staff who exhibit unhygienic practices.

11.23 The site should develop and implement a sickness reporting procedure, which identifies the actions to be taken should an individual fall ill either before or during his/her period of work. All staff should notify the manufacturing business in the event that they or a person they have been in contact with is suffering from or carrying an illness that could cause a food poisoning or food-borne illness outbreak before they commence work. The FSA Food Handlers: Fitness to Work—Regulatory Guidance and Best Practice Advice for Food Business Operators 2009, published in May 2009, identifies best practice in this area. Procedures should be in place should an individual fall ill at work in terms of reporting procedures, and these should form part of induction training. Procedures should also be implemented for managing any spillage of bodily fluid spillages, for example, vomiting and bleeding, within the production and storage areas. Bodily spill kits are available, which can be used and then disposed of in a way that minimises the risk of contamination. All contaminated products should be disposed of.

Agency Staff   11.24 The employment of agency staff should be in compliance with the legal requirements operating in the country where the manufacturing site is situated. In the UK, employment agencies that operate in food processing must hold a valid Gangmaster Licensing Authority (GLA) licence. The onus is on the manufacturer to check at routine intervals that the agencies' licence has not been revoked. In order to manage this effectively, a formal procedure should be in place with responsibilities defined as to who will coordinate and implement the procedure and, if required, appropriate corrective action. The training requirements outlined in this chapter apply as equally to agency staff as they do to employed staff. If training and medical records are held by the agency, they should be routinely verified as appropriate by the quality control manager or designate. If the agency is responsible for the laundering of protective clothing, then they should comply with all necessary controls (see 11.21h for more information).

Ethical Trading 11.25 It may be a requirement of retail customers as a prerequisite to supply
Initiative Base Code   that the manufacturer complies with the Ethical Trading Initiative (ETI) Base Code, SA 8000 or another third-party standard. It is best practice for manufacturers where this is a requirement to ensure that their suppliers comply with the relevant ethical trading standard(s) as well. It may also be a supply chain requirement for a manufacturer to be a member of the Supplier Ethical Data Exchange (SEDEX) as well as having a formal Sedex Members Ethical Trade Audit (SMETA).

12   PREMISES AND EQUIPMENT

Principle

Buildings should be located, designed, constructed, adapted and maintained to suit the operations carried out in them and to facilitate the protection of materials and products from contamination or deterioration. Equipment should be designed, constructed, adapted, located and maintained to suit the processes and products for which it is used and to facilitate protection of the materials handled from contamination or deterioration. Consideration should be given where appropriate to segregation of personnel and equipment from high- and low-risk food production. Physical separation of high- and low-risk food production and physical segregation of production of foods containing major allergens (see Chapter 4) should also be considered as part of the design process.

General

12.1   Depending on the products being handled, reference should be made to the detailed requirements in respect of premises and equipment in **EU Regulation (EC) No 852/2004 on the hygiene of foodstuffs, and EU Regulation (EC) No 853/2004 laying down specific hygiene rules for food of animal origin** (or with any regulations that may at any future time supersede these regulations). A site plan should be available that defines the location of the manufacturing unit in terms of buildings and external activities and services, for example, chemical storage, waste storage, raw material, packaging and production areas. The site plan(s) should also identify personnel flow around the site including internal and external access, staff facilities and pedestrian routes to work areas (see 12.18). Areas that may be subject to restricted access should also be highlighted (see 12.2). High-risk and low-risk product areas, where applicable, should be identified together with the production and process flow, routes for waste, quarantine areas and the movement of reworked products. The FSA publication *E. coli* O157—Control of cross-contamination: Guidance for food business operators and enforcement authorities (2011)[1] states that premises should be designed to ensure adequate physical separation; anything else 'will involve a shift towards greater uncertainty regarding the stringency of risk reduction that can be achieved'. Where there is dual use of equipment in manufacturing for both raw and high-risk products or low- and high-risk products, suitable procedural controls must be in place. The procedural controls must be validated and then consideration should be 'given to the monitoring and management arrangements required to ensure proper implementation of these procedures'.

Premises

12.2   Premises should be sited with due regard for the provision of services needed and to avoid contamination from adjacent activities. In existing premises, effective measures should be taken to avoid such

---

[1]http://www.food.gov.uk/business-industry/guidancenotes/hygguid/ecoliguide#.UFsKU42PXWE.

Food & Drink – Good Manufacturing Practice: A Guide to its Responsible Management, Sixth Edition.
The Institute of Food Science & Technology Trust Fund.
© 2013 John Wiley & Sons, Ltd. Published 2013 by John Wiley & Sons, Ltd.

contamination. Any measures introduced should be routinely reviewed to ensure that they remain effective. Examples include the provision of water, ice, compressed air, gas and air supply to the production area. The grounds surrounding the buildings should be maintained to minimise potential harbourage for pests that could be afforded by old pallets, packaging, waste, equipment and machinery or vegetation, which is growing up to the external walls of the building. Vegetation should be cleared on a regular basis, and there should be a minimum of 1-metre gap between vegetation and stored materials and buildings to prevent pest harbourage. Best practice would be to increase this distance further. Consideration should be given to adequate drainage in the yard and minimising pools of waste water through which vehicles may pass and then enter food manufacturing areas. The potential for flooding should also be considered especially with regard to sudden storm events. If the drainage system could not cope with such events then appropriate emergency measures should be available. External traffic areas should be suitably surfaced to prevent contamination or damage (through concussion) of product and packaging. This is especially important when glass product containers are used.

Building integrity should ensure that there are no access points for pests to gain entry or allow ingress of water through seepage or poorly designed guttering. Site security should be reviewed and a risk assessment documented with regard to the potential for malicious tampering or criminal activity with a view to contaminating the product. The need for restricted areas should be considered including the need for swipe card or keypad entry. The requirement for closed-circuit television (CCTV), security guards and/or fencing that fully encloses the site should also be considered as well as how staff, visitors and contractors are authorised to enter certain areas. Protocols for lone working should also be considered not only with regard to health and safety, but also with regard to product integrity and the potential for malicious tampering/sabotage. Guard dogs must only be allowed access to external areas and must be under the control of their handlers and not be allowed to run free. If guard dogs are used on the premises, then a risk assessment must be undertaken to determine the level of product risk and associated controls put in place. This must be documented.

12.3   Premises should provide sufficient space to suit the operations to be carried out, allow an efficient work flow and facilitate effective communication and supervision. The need to segregate high- and low-risk operations and those involving major allergens should also be considered. Physical segregation of materials may be required or the premises may need to be designed to ensure that personnel cannot transfer from one area to another during their work activities, for example, a raw and cooked meats factory, which requires personnel segregation. Segregation should consider the following: type and hazards associated with raw materials; in-process and finished products especially the control of allergens; the flow of product, rework, waste, packaging and materials, equipment (including maintenance tools), personnel, airflow, utilities and water; and transfer points between high- and low-risk areas. If physical segregation including barriers is used, then consideration should be given to personnel health and safety in the event of an emergency, for example, fire. Fire doors should therefore be alarmed or tamper evident so they cannot provide a personnel thoroughfare into and out of food areas.

The FSA publication *E. coli* O157—Control of cross-contamination: Guidance for food business operators and enforcement authorities (2011) paragraph 54 to 60 recommends that for extra protection against cross-contamination, a liquid hand wash that has disinfectant properties conforming to the European standards BS EN 1499:1997 is used.

When designing or redesigning the premises, the following may have to be taken into account:

- the site must comply with regulations that require the site to be approved or registered in terms of hygiene and premises and processing standards with the local government authority;
- availability or requirement for services such as electricity, power, sewerage, waste disposal, airflow, refrigeration plant, drainage;
- the requirements for effluent treatment prior to discharge, and provision for waste control including material segregation for recycling, where deemed necessary;
- the condition of external infrastructure including concrete and hardstanding and the potential for fugitive release of potential pollutants; condition of bunding and containment measures and whether they are suitable for the materials being stored;
- the need to define vehicle and pedestrian routes both internally and externally within the site (see 12.18);
- the availability or requirement for water, ice and/or steam within the process and its suitability for use. The source of the water, for example, mains, borehole (well), surface water or recycled water systems, should be considered and its ability to meet potable water standards (see Chapter 13). The requirement for, and control of, cooling water systems should also be considered and the safety procedures that are required especially with regard to *Legionella*. A water distribution plan should be available that includes all pipework and water-holding tanks as well as the type of water, that is, potable or recycled;
- the raw material/ingredient storage and types of storage required whether frozen, ambient or chilled;
- the retail crate/tray and bulk container (field tray, fruit bin, etc.) storage facilities especially if this is external. Procedures must be in place to ensure they are adequately stored to prevent contamination, especially if the product is loose and could come into direct contact with the trays, and that the trays are visually inspected before use;
- the requirement for deboxing–debagging areas for removal of external packaging before items gain entry into the production areas;

- the availability of designated in-process storage and the types of storage required whether frozen, ambient or chilled;
- the suitability of finished product storage prior to despatch and the types of storage required whether frozen, ambient or chilled;
- that storage areas should be of sufficient size to enable all operations to be carried out under appropriate conditions;
- the location of inspection and quality control stations within the premises;
- the location of label and packaging printing room (where this task is undertaken off-line);
- the material used for building fabrication and processing equipment. The presence of wood should be minimised and where possible eliminated;
- the production line and equipment layout to seek to maximise efficiency and minimise the risk of product contamination especially with regard to allergen control and the packing of identity preserved materials such as conventional, assured and/or organic food products;
- the facilities for equipment and premises cleaning and disinfection;
- the cleaning chemical and cleaning equipment storage areas;
- the design of product flow to minimise the risk of contamination and/or cross-contamination;
- the personnel flow in terms of personal health and safety, but also to minimise the risk of contamination and/or cross-contamination;
- the location and siting of hand-washing facilities especially the location and design of doors after hand washing has been completed in order to access production areas;
- the siting of equipment so that there is adequate access, especially for cleaning and servicing, around it;
- the requirement for charging battery-controlled equipment and the provision of suitable locations, for example, charging forklifts away from food environments;
- the restroom and personnel facilities;
- the requirement for protocol areas with barrier control in high-risk food premises; and
- the facilities for maintenance and equipment repairs.

For further information, refer to the Campden BRI Guidelines for the hygienic design, construction and layout of food processing factories (2003) Guideline G39 ISBN 0 905942 57 4.

12.4   All premises including processing areas, laboratories, stores, passageways and external surroundings should be maintained in a clean and tidy condition.

12.5   Premises must be constructed and maintained with the object of protecting against the entrance and harbouring of vermin, birds, pests and pets (see also 12.11 and Chapter 15, Infestation Control).

12.6   Premises should be maintained in a good state of repair. The condition of buildings should be reviewed regularly, and repairs effected where necessary. Special care should be exercised to ensure that building materials and construction, repair or maintenance operations are not allowed to affect adversely product quality or integrity. Where appropriate to the type of manufacture, positive air pressure systems should be installed. Buildings should be effectively lit and ventilated to minimise dust and prevent condensation, with air control facilities (including temperature, humidity and filtration), appropriate both to the operations undertaken within them and to the external environment.

12.7   There should be sufficient intensity of light to aid the activities being undertaken, for example, product inspection and cleaning. Areas should be lit to enable personnel to work safety especially where there is no external light entering the building or personnel work over a 24-hour period. All light appliances should be protected by shatterproof plastic diffusers or sleeve covers. Where this is not possible, a fine metal mesh screen must be used. The lights should be subject to brittle material control procedures (see 12.36).

12.8   Working conditions (e.g. temperature, humidity, noise levels) should be such that there is no adverse effect on the product, either directly or indirectly via the operator.

12.9   Fans should be sited to avoid contamination hazards caused by either intake of noxious solids, vapours or gases, or exhaust of materials that could contaminate other products, and with due regard for the local environment and the avoidance of nuisance. Nuisance could include odour, noise or dust emissions.

12.10   Air supply and extraction trunking should not introduce contaminants into products. For dry food products, dust extraction equipment may need to be installed. Ventilation and extraction systems must be suitably designed to meet processing requirements in terms of preventing condensation and pest ingress, and managing dust effectively. Systems should be sited to minimise contamination of the product or process. Air that comes into contact with the product should be filtered. The air-handling system must be designed to address:

• the degree of variance in ambient air, for example, temperature, humidity, level of dust and particulates;
• the process conditions in terms of temperature, humidity, air containment, air extraction requirements, levels of dust and particulates in the process air; and
• the air filtration cycle in terms of positive pressure, volume of air supply and air balancing.

Air filtration equipment must be adequately maintained and regularly inspected, and air quality routinely monitored. A risk assessment should be undertaken to determine the need for positive air pressure between high- and low-risk areas, the filter size/grade required to control airborne contamination and the potential for contamination, for example, the product residency time in that area. If air socks are used, they must be inspected, cleaned and maintained at a designated frequency based on risk assessment. For further details, consult the particulate air filters for general ventilation: Determination of the filtration performance standard BS EN 779:2002 and the Campden BRI Guidelines on air quality standards for the food industry (second edition) 2005; Guideline G12 ISBN 0 905942 73 6.

12.11 Floor and wall surfaces, and all surfaces (including surfaces of equipment) in contact with food must be maintained in a sound condition, and they must be easy to clean and, where necessary, disinfect. Walls should be sound and finished with a smooth impervious and easily cleaned surface. Where walls are in areas of high traffic movement and there is the potential for damage, crash barriers should be installed. Floors in manufacturing areas should be made of impervious materials, laid to an even surface and free from cracks and open joints. They should be of adequate construction and material for the wear and tear and conditions of manufacture encountered. Floors should be designed so that liquid does not collect in certain areas, and the fall of the floor should be such that any water or waste product travels easily to a suitable drain.

Ceilings should be so constructed and finished that they can be maintained in a clean condition. Voids above false ceilings should be regularly cleaned and inspected so that they do not provide harbourage to pests. It is essential to ensure effective seals to walls and floor. The coving of junctions between walls, floors and ceilings in critical areas is recommended.

Processing areas should be designed without windows. However where windows are present in production and storage areas, they should be of toughened glass or plastic, protected against breakage, adequately screened to prevent pest entry and secured, and with ledges sloped away from the glazing. Plastic film can be adhered to internal window surfaces as an extra control in the event of damage or breakage.

Doors, dock levellers and door frames should be of impervious, non-corrodible material; smooth, crevice-free and easily cleanable; and suitably protected to prevent ingress of pests when opened (see Chapter 15, Infestation Control). Doors must be monitored to ensure that they remain close-fitting and provide adequate pest-proofing. External doors, especially pedestrian doors should be designed to be automatic closing doors, and if deemed appropriate, air curtains should be installed. Consideration should be given to the design of doors as to whether they are suitable for food preparation areas especially where high-risk products are produced. The design of personnel door should minimise handles that can be a source of cross-contamination between personnel, have kick plates to prevent damage when opening, be seamless where possible to prevent bacterial harbourage and be water resistant in areas of water use. Consideration should be given to whether sliding doors or roller doors are used. Roller doors are in contact with the floor and on lifting, depending on the food environment, can drip material onto product or equipment as it passes underneath. If roller doors are used, a risk assessment should be undertaken to determine the degree of risk of product contamination with such doors being in place. If sliding doors are used, it should be determined if there is any health and safety risk to personnel if they are working in close proximity to the doors when they are opening or closing. Automatic doors should be open for an appropriate time, which allows access but prevents ingress of pests and other potential product contaminants. Where strip curtains or similar designs are used, they must be maintained so that they are effective against pest ingress, be cleaned at a prescribed frequency and otherwise controlled to prevent product contamination. Training of staff should include effective door management in order to prevent product contamination, ingress of pests and loss of temperature control. Materials should be chosen so as to avoid tainting or otherwise contaminating food materials. Painted surfaces are not ideal because flaking may occur, which could pose a potential food contamination risk. The paint if used should comply with BS 4310:1968, the standard for low-lead paints. Specialist paint is available for the food industry and advice should be sought.

12.12 Pipework, suitably protected light fittings, ventilation points and other services in manufacturing areas should be of a material suitable for purpose and appropriate to the area where the services are located, afford cleaning and disinfection if required and be sited to avoid creating recesses that are difficult to clean. Services should preferably run outside the processing areas. They should be sealed into any walls and partitions through which they pass to prevent pest ingress. Intake points where product is received from external deliveries through pipework, or other means, to internal tanks must be suitably designed and operated to be easy to clean, and where necessary disinfect, and prevent ingress or harbourage for pests. End caps should be fitted to all external points, or they should be otherwise enclosed, to prevent contamination and/or malicious tampering. When not in use, they should ideally be locked off to prevent entry. Fabrication joints must be sealed and not pose a product risk. The type of welding should be appropriate for the product being manufactured. Pipework should be checked on a regular basis to ensure full integrity and any potential for harbourage of harmful bacteria is minimised. Pipework should also be assessed for the potential build-up of biofilms. If identified, appropriate action should be taken.

12.13 Drains should be of adequate size and should have trapped gullies and proper ventilation. Any open channels should be shallow to facilitate cleaning. Machinery and sinks should be sited to ensure that process and other waste water is discharged directly to drain. Drains must flow from high- to low-risk areas, and a system must be in place to prevent back flow of liquid and air. Drains from laboratories should also be considered to determine the potential for product contamination and where required appropriate action taken ideally at the design stage to minimise risk. A drainage plan should be in place for both internal and external drainage on the site. Drainage surveys, including the use of cameras, may need to be undertaken to determine the condition of the infrastructure, potential for seepage, flow of foul (dirty) and surface water drainage and whether the drainage links to a treatment process prior to discharge.

12.14 An adequate number of flush lavatories must be available and connected to an effective drainage system. Lavatories must not lead directly into rooms in which food is handled or stored (see 11.21c).

12.15 Protection from the weather should be provided for receiving and despatch areas, and for materials or products in transit. External storage is not recommended, but where external storage is necessary, procedures must be in place to minimise deterioration and the risk of contamination.

12.16 Where raw materials or packaging materials arrive in external packaging, a separate deboxing or debagging area should be provided where the packaging may be removed before the materials enter the production area.

12.17 Waste material should not be allowed to accumulate. It should be collected in suitably constructed and identified receptacles for removal to collection points outside the buildings, and be disposed of at regular and frequent intervals. Disposal of printed packaging materials or raw materials and rejected products should be carefully controlled (see Chapter 22).

12.18 Design of premises should provide separate routes of entry and movement for vehicles and personnel. Designated walkways should be marked in internal and external areas. Where walkways or steps cross over production lines, storage areas or workstations, they must be fitted with backplates, and/or be suitably enclosed to prevent contamination of the product, packaging or people. The material utilised to make the stairs must be suitably controlled to prevent contamination. Ideally, wood would not be used in such areas. If wood has been used, an appropriate risk assessment must have taken place to determine the degree of risk. The use of ladders or scissor lifts in production and processing areas should also be considered and appropriate action taken to minimise the potential for contamination of the area and the food itself. When designing premises, it must be considered that manufacturing areas should not be used as a general right of way for personnel or materials, or for storage. Sufficient storage space must be available so that material storage at no time prevents or limit access to fire exits. This would not exclude materials temporarily standing for a brief time in transit from storage to a production area. The space required for temporary storage must also be considered in premises design.

12.19 All operations should be carried out in such a way that the risk of contamination of one product or material by another is minimised.

12.20 There should be documented cleaning procedures and schedules for external, manufacturing and storage areas (see Chapter 14).

12.21 Vacuum or wet cleaning methods are to be preferred. Compressed air, hoses, pressure cleaners, brooms and brushes should be used with care, so as not to incur the risk of product contamination. Where compressed air is used, the potential for air contamination should be considered and appropriate action taken.

12.22 Surfaces in contact with food should be inert to the food under the conditions of use and should not yield substances which might migrate or be absorbed into the food. A certificate or other written declaration of compliance should be held for all food contact surfaces to demonstrate that the materials are "food grade" and comply with **Regulation (EC) No 1935/2004 on materials and articles intended to come into contact with food and repealing Directives 80/590/ EEC and 89/109/EEC and Regulation (EC) No. 2023/2006 on good manufacturing practice for materials and articles intended to come into contact with food and Regulation (EC) No. 450/2009 on active and intelligent materials and articles intended to come into contact with food as well as The Materials and Articles in Contact with Food (England) Regulations 2010** or its equivalent. **Directive 97/48/EC** lays down the basic rules necessary for testing migration of the constituents of plastic materials and articles intended to come into contact with foodstuffs and all material testing must comply with the requirements laid down in this directive or as subsequently superseded. The list of groups of materials and articles which could be covered by specific measures includes not only food grade material used in equipment design, but the following seventeen materials and processing aids:
   1. active and intelligent materials and articles

   2. adhesives

   3. ceramics

   4. cork

   5. rubbers

   6. glass

   7. ion-exchange resins

   8. metals and alloys

   9. paper and board

   10. plastics

   11. printing inks

   12. regenerated cellulose

13. silicones

14. textiles

15. varnishes and coatings

16. waxes

17. wood.

Compliance with this legislative requirement should be a prerequisite that is built into the process design phase of a new process line or piece of equipment. Surfaces should be durable and able to sustain the activities undertaken without cracking or deterioration, for example, inspection tables, conveyor belts and elevators. Where food contact surfaces are liable to wear, they should be of a contrasting colour to the food and raw materials used in manufacturing and should be inspected regularly for evidence of damage.

12.23   Surfaces in contact with food should be microbiologically cleanable, smooth and non-porous so that particles are not caught in microscopic surface crevices and become difficult to dislodge. Welds should be continuous, and all welds and joints should be smooth and impervious. Routine swabbing should be undertaken to determine the efficiency of cleaning (see 12.12).

12.24   Surfaces in contact with food should be readily accessible for manual cleaning or if not readily accessible, then easily dismantled for manual cleaning, or if clean-in-place (CIP) techniques are used, it should be demonstrated that the results achieved without disassembly are the equivalent of those obtained with disassembly and manual cleaning.

12.25   Interior surfaces in contact with food should be so arranged that the equipment is self-emptying or self-draining.

Equipment   12.26   All equipment should be purchased in accordance with specified requirements. It should be suitably tested and commissioned before the line is approved for full-scale production. This could include small production runs with a microbiological sampling plan and shelf-life testing to validate key processes (see 3.11). Equipment should be cleaned and serviced before use and any faults rectified. During commissioning, equipment should be tested to ensure that it meets the purchasing specification and is fit for its intended purpose and can be effectively cleaned and maintained.

12.27   Equipment should be so arranged as to protect the contents from external contamination, and should not endanger a product through contamination from leaking glands, condensate, lubricant drips and the like, or through inappropriate modifications or adaptations.

12.28   Exterior surfaces of equipment not in contact with food should be so arranged to prevent harbouring of soils, microorganisms or pests in and on the equipment, floors, walls and supports.

12.29 Equipment should be designed and constructed to allow efficient cleaning and maintenance. There should be detailed written instructions for cleaning and disinfection. Specified materials, methods, safety precautions and suitable facilities should be provided (see Chapter 14).

12.30 Plant and equipment should be cleaned and serviced immediately after use. Any faults should be recorded. Missing parts, such as nuts, bolts, springs and clips, should be reported immediately to the quality control manager or designate.

Maintenance 12.31 The structure of the building including walls, floors, ceilings and doors should be maintained in a sound condition (see 12.11). Plant and equipment should be checked for cleanliness and integrity before every use and to this end should be designed with sound, secure, quick-release systems for inspection and disassembly. Appropriate precautions for ventilating fumes from power-driven equipment, heaters and so on should be taken with consideration for both product safety and personnel health and safety. Temporary repairs should be controlled to ensure they do not impact on the safety or legality of product. Temporary engineering measures such as by-pass pipework should be permanently repaired as soon as feasible, subject to processing constraints.

12.32 A documented maintenance procedure should be developed, which addresses both preventive and responsive maintenance. The procedure should be based on risk assessment and ensure that all servicing and breakdown work undertaken on equipment is carried out, ensuring that food safety, legality and quality requirements are achieved and that any potential risk to the product is minimised. Preventive maintenance should be considered for all equipment and components that contribute to product safety. The frequency of servicing should be determined by the maintenance manager using factors such as manufacturer's recommendations, equipment and plant history in terms of failure or breakdowns, proposed hours of use and production requirements. This should be documented on a planned maintenance schedule. The planned maintenance schedule needs to be reviewed on a routine basis and amended as required. It should include all equipment and plant in the manufacturing unit and storage areas that ensures that the environment and the food products are manufactured to comply with legislative requirements. The fabric, plant, equipment and machinery should be serviced/maintained according to the maintenance schedule and any actions required documented in a machine log or other such internal document. This will then provide a service history for the piece of equipment plant or area. The planned maintenance schedule may need to be updated in the event that hygiene audits or other internal audits identify problems with poor maintenance, for example, cracking floor joints, peeling or flaking paint, damage or wear. Maintenance and machine servicing contractors must be made aware of the organisation's maintenance procedure before they commence work, and particular attention must also be paid to personnel hygiene, premises hygiene and food contamination procedures that are in force to ensure the contractors' compliance at all times. Waste control procedures must also be formally agreed with contractors to ensure adequate control and compliance with relevant legislation.

12.33   Responsive maintenance may be required during production following identification of a machine fault. Minor adjustments could be carried out without interruption to the production line where the appropriate personnel have deemed that it does not present a risk to the product. Major repairs or adjustments would require the production line/area to be closed down and if necessary be screened off. Any parts that are removed should be logged and accounted for by maintenance personnel prior to the line starting up again or the area go back into production. A cleaning procedure should be defined for the activities to be undertaken before a machine/production line is deemed suitable to go back into production. A hygiene clearance/sign back to production record should be maintained detailing, as applicable, the equipment, brief details of the task undertaken and any parts that have been removed or replaced. The record should be signed off by the maintenance personnel concerned as complete and by production personnel to authorise recommencement of production. All maintenance and servicing activities must be undertaken by competent staff (including contractors) who can demonstrate by training or other means that they are competent to undertake the task required and are aware of and comply with internal procedures.

12.34   Consideration should be given to the potential for cross-contamination between low- and high-risk areas by maintenance personnel, their tools and equipment. Consideration should also be given to managing the contents of engineers' tool boxes to ensure that all items taken into the production area are removed and/or accounted for before production can commence/recommence. Tools must present no contamination risk to the product and must be clean, and well maintained. There should be information available that demonstrates that all materials, oils and lubricants utilised during maintenance are food grade and do not present a risk to product safety.

12.35   All engineering workstations/workshop must be controlled so that they do not present a risk to product safety or legality. They must be maintained in a suitable hygiene status, and controls should be put in place such as the use of swarf mats and waste control to reduce risk to product safety.

Brittle Material   12.36   Wherever possible, the use of glass should be avoided and suitable
Procedures   alternatives sought. There are three types of glass: packaging; fixtures and fittings such as glass windows, light bulbs and dials; and imported glass such as spectacles, watches and drink bottles. Windows should be secured and laminated to contain any glass in the event of a breakage. Light bulbs and fluorescent tubes should be completely covered to contain any pieces of glass in the event of shattering, or be made of shatter-resistant material. A register should be maintained, and updated as necessary, of all glass, ceramic and

hard plastic items in the production area or items that are deemed to present a product contamination risk. A risk assessment should be undertaken to determine the frequency of inspection of all items whether each shift, daily, weekly or other prescribed interval. Items should be regularly inspected and the results of the inspection recorded with any required corrective action. This corrective action must be followed up to ensure completion and the removal of a potential product contamination risk. The use of glass and ceramic containers should be prohibited in all production areas. If glass or ceramic containers are used for the final product, then breakage procedures should be in place, for example, delivery, storage, transfer, visual inspection on-line and filler breakage procedures. A written procedure must be in place for the replacement of brittle items detailing the measures that are in place to prevent breakage and to minimise product contamination.

Consideration should be given to the use of equipment/items that are made with brittle materials being present in offices that open directly into storage and production areas. A risk assessment should be undertaken and recorded, which outlines the potential for contamination and the controls in place to minimise the risk of occurrence.

12.37   Any breakage of glass lenses in spectacles should be treated as a glass breakage incident and must be controlled by relevant procedures. Procedures should also be in place in the event that a contact lens is lost.

12.38   All personnel must report immediately to their line supervisor or management any broken or damaged glass/hard plastic or ceramic items. This applies to any location within the production unit or site. Any incident of broken glass or hard plastic components on processing equipment or other glass, ceramic or hard plastic breakage that could in any way have affected the product should result in production being stopped and the product immediately held. Any product/part-processed product or ingredients that could be contaminated must be rejected and a thorough search of the area made before any more product is packed/processed. The area and all equipment within a designated (e.g. 10-metre) radius of the breakage incident must be isolated immediately and thoroughly searched for fragments.

12.39   All personnel in the area at the time of the incident should have their clothing and the soles of their footwear checked for glass, ceramic or hard plastic fragments before leaving the area. All remaining fragments should be removed immediately by using the dedicated equipment for removing glass and disposed of carefully. Colour-coded equipment is often used, for example, red bins and red brushes and equipment, which is designated 'GLASS ONLY'. Production equipment that may have been affected must be dismantled for in-depth inspection and cleaning. Broken or cracked windows should be removed from the outside, with heavy-duty polythene sheeting taped to the inside of the production area to prevent glass spillages. The glass must be replaced with a suitable alternative. Where practical, fragments should be pieced back together to try and account for all the glass, ceramic or hard plastic. A sample fragment should be retained for reference in the event of a subsequent customer complaint and for further analysis if necessary.

12.40   The area must be closely inspected again after cleaning and when the area has been declared free of glass, ceramic or hard plastic, formal documentation should be completed and signed off by the appropriate manager to formally clear the area for recommencement of production.

12.41   For glass/hard plastic or ceramic breakages in areas remote from the production and storage area, for example, offices and restrooms, a thorough inspection of the area should still be made by appropriate personnel in order to assess the potential risk to product. The breakage procedure must be followed as stated above.

12.42   Where other brittle items present a potential contamination risk to the manufacturing process, they should be addressed by a similar procedure.

Knife/Scissors/Blade   12.43   Knives, scissors and blades would generally be used for the following
Procedures   activities:

- opening outer packaging;
- cutting through shrink wrap, cardboard;
- cutting packaging fastenings; and
- food preparation.

Food contact equipment should be identified and kept separate from equipment used for other activities, for example, designated colours for specific activities or departments, for example, goods inwards, production or quality control. Knives (and all other sharp equipment) should be company issued and only be allowed in defined areas. Special protected blades are ideal, which are designed for safety and do not have an exposed blade. Knives should be controlled by suitable documented procedures including the signing out of equipment to an individual and the signing back in again at the end of the production shift on the equipment's return. Where possible, knives should be chained or otherwise securely fastened in a location to prevent their unintentional loss. Knives should be uniquely identified so that they can be assigned to individuals. On return, they should be checked to ensure there are no signs of damage and the blade is intact. Where appropriate, knives should be sterilised to minimise cross-contamination. In the event that a knife is 'lost' or unaccounted for, the management of incidents procedure should be implemented (see Chapter 19).

Consideration should be given to monitoring the condition of knives that form an integral part of machinery. They should be inspected on a regular basis for integrity, and this inspection should be recorded. Consideration should also be given to implementing metal detection as a measure to ensure the potential for metal contamination is adequately controlled.

Where knives are used in the preparation of food items during manufacture, for example, vegetable and meat trimming, due consideration should be given to implementing an effective disinfection programme. In order for such treatments to be effective, a cleaning procedure must be in place to remove soil and organic matter; otherwise disinfection by hot water above 82°C, or the use of chemical disinfectants will not be effective. Routinely the effectiveness of the cleaning and disinfection process should be verified by swabbing of the knives. It should be noted that spores produced by pathogenic organisms may survive disinfection and present a food safety risk. This risk is both process and product specific.

13   WATER SUPPLY

Principle

An adequate supply of potable water should be available. Provision should be made for appropriate facilities for temperature control, storage and distribution of water. Dead legs should be eliminated in water systems, and if cooling towers and/or showering facilities are used, consideration should be given to monitoring for *Legionella*.

General

13.1   **The EU Regulation (EC) No 852/2004 on the hygiene of food-stuffs** requires an adequate supply of potable water. This potable water must be used whenever necessary to ensure foodstuffs are not contaminated. Ice must be made from potable water to ensure food-stuffs are not contaminated. It must be made, handled and stored under conditions that protect it from all contamination. Steam used directly in contact with food must not contain any substance, which presents a hazard to health, or is likely to contaminate the product. **The EU Regulation (EC) No 852/2004 on the hygiene of food-stuffs** states that water unfit for drinking used for the generation of steam, refrigeration, fire control and other similar purposes not relating to food, must be conducted in separate systems, readily identifiable and having no connection with, nor any possibility of reflux into, the potable water system. A water distribution map should be available and appropriate controls should be in place (see 3.9, 9.1e and 12.3). A cleaning schedule should also be in place for the water distribution system, which had been validated, monitored and is routinely verified. The water distribution map can be colour coded to aid differentiation of water systems.

13.2   Water supplies must be potable and be drawn from mains water or otherwise treated to meet potability standards. Routine microbiological sampling should be undertaken to ensure that the water meets these standards and appropriate action undertaken if results indicate quality standards have not been met. Potable water should be as a minimum standard as specified in the latest edition of World Health Organisation (WHO) Guidelines for Drinking Water (currently 4th edition published 2011). Potable water used for manufacturing purposes, including that used in making up products or likely to come into direct contact with the product, must be of potable quality and free from:

(a) any substances in quantities likely to cause harm to health;
(b) any substance at levels capable of causing accelerated internal corrosion of metallic containers and closures causing taints;
(c) harmful microorganisms.

In addition, coliforms should not be detectable in 100 mL in 95% of the samples taken nor in any two consecutive samples. Aerobic plate

Food & Drink – Good Manufacturing Practice: A Guide to its Responsible Management, Sixth Edition.
The Institute of Food Science & Technology Trust Fund.
© 2013 John Wiley & Sons, Ltd. Published 2013 by John Wiley & Sons, Ltd.

counts (APCs) should not exceed 100 organisms/ml after incubation for 5 days at 25°C. Checks for presumptive coliforms should be made at least monthly and those for APCs at least weekly. To maintain the above standards, the water should, if necessary, be chlorinated or otherwise adequately treated. If water is treated, there should be a procedure in place to ensure that the treatment of water is adequately controlled to prevent food contamination. Monitoring must be adequate to control any trends towards a critical limit especially if chlorine control is deemed within the hazard analysis critical control point (HACCP) plan as a critical control point (CCP) as described in Chapter 3. It is important in this context to differentiate between determining critical limits for both free and total chlorine. All critical limits should be validated before the HACCP plan is implemented. Corrective action procedures must be in place if target levels or indeed critical limits are breached including disposition of product. Public water providers can supply certificates of potability, and these should be retained by the manufacturer and reviewed for any trends that could impact on water suitability. It is important to ensure that the certificates supplied by public water providers are not too generic across a wide water distribution network, but actually relate specifically to the local water distribution system from which the manufacturer receives water.

Although the water may be delivered to the manufacturing site at a potable standard, this does not mean that at the point of use, it is still at the selfsame standard. Consideration should be given to the number of individual usage points and therefore the number of sampling points and frequency of sampling required to verify consistent standards. A risk assessment should be undertaken to determine the level of risk at the point of water use with regard to food safety and quality. The nature of the process is also important and whether the product comes into direct contact with water, or water is a product ingredient. A further consideration is if the product comes into direct contact with the equipment, conveyors and so on as this will increase the risk of contamination. Further factors to be reviewed include, but are not limited to, the original source of the water; procedures for storage of water on-site; materials used in water storage tanks and transfer pipework and whether they predispose the system to the development of biofilms, or contamination of the water; the water treatment process that is undertaken prior to use especially if the treatment itself could cause a potential food safety hazard; and the management of waste water.

13.3   Water that is recycled for reuse should be treated so that it does not pose a potential contamination risk to the premises or product. The treatment process should be routinely monitored to determine its effectiveness. If there are both potable and recycled water systems in the manufacturing unit, they should be suitably identified to avoid confusion so that operatives are aware at all times whether water at an outlet point is potable/non-potable. Water systems may be differentiated using colour coding on taps, coloured hoses or designated water site mapping as previously described.

Boil Water Advice   13.4   In the event of microbiological contamination of mains water supplies, water utilities in the area(s) concerned would, in most instances, issue advice to boil water before use. The management team with advice from the quality control manager would have to determine whether it was possible to continue manufacture of food products under such circumstances. Further details can be found in the following publications:

- Institute of Food Science & Technology (IFST) Advisory Statement on Contamination of Water Supplies: Boil Water Advice of January 2004
- Water Quality for the Food Industry: Management and Microbiological Issues (2000). Guideline No. G27. Campden BRI. ISBN: 0 905942 31 0
- The **EU Council Directive 98/83/EC** on the quality of drinking water for human consumption as amended by regulations **1882/2003/EC** and **596/2009EC**
- see http://ec.europa.eu/environment/water/water-drink/index_en.html.

Water Sampling   13.5   Water should be routinely sampled both at the point of entry to the
Procedure   site and at various points around the premises. This will give a good indication of the loading on entry to the site and whether the water system is increasing the bacterial loading of the water. If this is found to be the case, then remedial action will need to be taken to reduce the bacterial loading. The quality control manager should develop a water sampling protocol based on risk assessment. The risk assessment should consider the frequency of sampling required for both microbiological and chemical analysis of water to verify that it meets minimum potable standards. The risk assessment should also consider the proportion of the finished product that constitutes water added during the manufacturing process such as in the case of diluting fruit juice concentrate in order to pack single strength juice. The inclusion of water, where water forms a constituent of the product, for instance carbonated drinks or meat preparations where the meat contains added water, means that the frequency of water sampling as defined by risk assessment may well be much higher than when water is utilised only for cleaning and disinfection activities. Consideration of factors such as nitrate, arsenic and pesticide levels in water should be considered where the water source is borehole (well) or surface water so that the source can be initially validated and then routinely verified. The use of recycled water needs to be considered in the risk assessment and appropriate frequency of sampling determined to ensure that water treatment has been adequate. Ice should also be sampled and tested both ice manufactured on-site and ice purchased and delivered to the site.

             13.6   Samples may be analysed by an in-house or a third-party laboratory. If a third-party laboratory is used, it should be formally approved (see Chapters 16 and 29). Sterile minimum 500-mL sample bottles should be used, and water should be run for a period of time before the sample is taken. Any contamination of the sample bottle and the lid during sampling should be avoided. Ideally samples should be under adequate temperature control during storage and transit and be with the laboratory within 4–8 hours of being taken. The samples should be kept at a temperature of between 2 and 10°C while in transit and storage and ideally in the dark. This may require the use of a hygienic cool box and/or ice packs.

13.7   If the water is chlorinated, sodium thiosulphate needs to be added to the sample bottle to neutralise any residual free chlorine, so it is important to agree with the laboratory the type of sample bottles required.

13.8   When the results are received from the laboratory, they should be reviewed to check for compliance with the specification. If the results are out of specification, this must be referred to the quality control manager and appropriate corrective action taken. This could include a review of product status, the potential for product contamination and the need for a product recall.

14 CLEANING AND SANITATION

Cleaning and sanitisation programmes are a key prerequisite to ensure the adoption of good manufacturing practice (GMP). Cleaning will not be effective unless it is fully supported by management. The role of management is to define the hygiene standards required and to communicate these effectively to staff, usually by means of a comprehensive cleaning schedule and associated task procedures. Management must demonstrate commitment by providing the appropriate means, that is, the equipment, the chemicals and the training for staff.

General

14.1 **EU Regulation (EC) No 852/2004** on the hygiene of foodstuffs states that:
1. Food premises are to be kept clean and maintained in good repair and condition.

2. The layout, design, construction, siting and size of food premises are to:

   (a) permit adequate maintenance, cleaning and/or disinfection, avoid or minimise air-borne contamination, and provide adequate working space to allow for the hygienic performance of all operations;
   (b) be such as to protect against the accumulation of dirt, contact with toxic materials, the shedding of particles into food and the formation of condensation or undesirable mould on surfaces;
   (c) permit good food hygiene practices, including protection against contamination and, in particular, pest control.

14.2 In the UK, Industry Guides on Good Hygienic Practice were developed by many food sectors to give advice on how to comply with the requirements of the former **Food Safety (General Food Hygiene) Regulations 1995** and the **Food Safety (Temperature Control) Regulations 1995** (see 11.15).

14.3 Cleaning procedures must not only be consistent with Food Hygiene legislation but also with the requirements of Environmental and Health and Safety (see Chapters 30 and 31) legislation and undertaken to minimise the risk of contamination.

Cleaning Schedule/ Procedures

14.4 Cleaning schedules coordinate cleaning activities and are a means of information transfer between management and staff. They should be established for external areas, premises, equipment, plant and services. They should also be developed for transport vehicles in the distribution supply chain. Cleaning schedules should be designed by individuals who have the appropriate competence and knowledge to

Food & Drink – Good Manufacturing Practice: A Guide to its Responsible Management, Sixth Edition.
The Institute of Food Science & Technology Trust Fund.
© 2013 John Wiley & Sons, Ltd. Published 2013 by John Wiley & Sons, Ltd.

develop an effective cleaning and disinfection regime. The site hygiene plan and associated cleaning schedule should be based on a risk assessment approach (see 14.8). Documentation such as cleaning schedules and associated task procedures must be clear and unambiguous. Consideration must be given to language skills, numeracy and literacy of staff. Cleaning schedules and associated task procedures must state as appropriate:

- what area or equipment is to be cleaned;
- who is to undertake the task (i.e. responsibility);
- when it is to be cleaned (i.e. how often);
- how it is to be cleaned (i.e. work instruction or task procedure including how the equipment should be dismantled and reassembled);
- the time duration of the cleaning task especially the required contact time, that is, the time that the cleaning chemical has to be in contact with the surface;
- the materials to be used, that is, cleaning chemicals and cleaning equipment (see 14.7);
- dilution rates and whether the chemicals are pre-diluted or need to be diluted by staff to the correct rate;
- the requirements for rinsing between chemicals and if required the protocol for the pH testing of rinse water;
- the cleaning standard expected in terms of either visual standard or if microbiological or adenosine triphosphate (ATP) sampling is undertaken the standards to be reached;
- the aspects that need to be addressed to ensure operator safety and the personal protective equipment (PPE)/clothing, which may need to be worn either when handling concentrated chemical and/ or when completing cleaning tasks;
- the health and safety of staff and the environmental concerns in the event of a spillage of chemicals and how such spillages should be contained and cleaned up;
- who should be contacted if there is a problem during the cleaning task;
- which cleaning records need to be completed and by whom; and
- who is responsible for monitoring the task to ensure it has been effective and that the records are complete to verify this action.

14.5   Cleaning schedules may be designed to address daily, weekly, monthly, quarterly, six-monthly and annual tasks separately or all on one format. Cleaning may be verified by the quality control function as part of a routine site hygiene audit, or it may be verified as part of production procedures by a hygiene team leader or equivalent.

14.6   The cleaning schedule needs to ensure effective, economic cleaning. Equipment should be installed so as to facilitate effective cleaning. The practices used will vary with the size of premises and number of staff involved. Cleaning schedules should be in place to ensure that external areas are kept clean and free from rubbish. Particular attention should be paid to waste storage areas. The cleaning requirements for chill and cold stores, road vehicles and shipping and air freight containers need to be considered. Cleaning schedules need to be developed with appropriate protocols to minimise the risk of product contamination and determine the effectiveness of cleaning.

The quality control manager or designate must determine a site hygiene plan, otherwise termed cleaning and disinfection protocol that is appropriate to the business and the products manufactured especially in terms of high- and low-care requirements. Consideration should be given to the generation of aerosols that could contaminate nearby surfaces, packaging, ingredients or products. The control of allergenic materials and the disposal of cleaning equipment that may have come into contact with allergenic materials must also be considered.

The implementation of the cleaning schedule will only be effective if there are sufficient resources available in terms of time, trained personnel, suitable chemicals and cleaning equipment. Consideration should be given to the level of cleaning, for example, routine cleaning at one level and then further deep cleaning programmes at designated intervals. Deep cleaning may need to be undertaken during periods of production shutdown and non-production, and engineering and maintenance resource may also need to be programmed where equipment dismantling is required. If equipment is dismantled, appropriate controls should be in place to ensure that machine parts are not placed directly onto the floor. Controls include, but are not limited to, tables, designated racking and so forth. A hygiene clearance/sign back procedure should be in place (see 12.33). The cleaning schedule and associated task procedures must be validated before implementation and monitored by staff to ensure it is sufficient to deliver food safety objectives. Verification activities must take place over and above monitoring to demonstrate that the system as a whole is functioning effectively (see 14.13).

Site Hygiene Plan 14.7 The quality control manager should develop and validate a site hygiene plan and associated procedures for the site. The site hygiene plan should be reviewed at regular intervals to determine its appropriateness and effectiveness in ensuring a hygienic manufacturing site and minimising the risk of potential product contamination. The Food Standards Agency (FSA) publication *E. coli* O157—Control of cross-contamination: Guidance for food business operators and enforcement authorities (2011) stresses the importance of not only validating the hazard analysis critical control point (HACCP) plan, but also ensuring that disinfectants are purchased and used in compliance with validated dilution levels and contact times. The guidance states that the use of disinfectants or sanitisers that meet BS EN 1276:1997 or BS EN 13697:2001 should be considered. In order to promote efficacy they must be applied to visibly clean surfaces and be used "strictly in accordance with the manufacturer's instructions relating to proper dilution of the chemical, the effective temperature range and the necessary contact time. Since effective chemical disinfection can only be achieved on visibly clean surfaces, a cleaning stage is required first." Validation of cleaning and disinfection processes is especially critical if dual use of equipment and machinery for slicing, mincing or vacuum packing of raw and ready-to-eat foods forms part of the manufacturing process. Where possible, there should be designated machinery to prevent this risk from occurring. Where this is not possible, effective disinfection is a key prerequisite to ensuring food safety. Validation of cleaning methods should also assess the potential for cross-contamination by cleaning equipment. The guidance further requires that food business operators (FBOs) should "ensure that they are using the appropriate disinfectant products by confirming with their suppliers that the products they are using meet, as a minimum, the specifications of these standards. This information may also be obtained from the label of the product, or by contacting the manufacturer directly."[1] Records should be maintained of all validation and re-validation activities.

14.8   Risk assessment must be undertaken, which is specific to the manufacturing unit to determine the level of cleaning and disinfection required. The site hygiene plan should be developed to address the following points as appropriate to the site and the type of product manufactured:

(a) the food products being manufactured will influence the requirements for those products in terms of cleaning and disinfection. Low-risk products may not require a full disinfection process to take place especially where the product is not in direct contact with the equipment, whereas a high-risk product will require both cleaning and disinfection to be undertaken at the food premises;

(b) whether a clean-down process is required between different products during a production run or the cleaning process will be undertaken at the start and end of the production period (i.e. daily, per shift);

(c) who will undertake the cleaning – will there be a designated hygiene team or will the production staff undertake the cleaning at the end of production or intervals as determined;

(d) if a hygiene team is used, will they be internal staff or will cleaning be outsourced to a cleaning contractor. Cleaning contractors should be approved as per the supplier approval procedure (see Chapter 16);

(e) the budget for cleaning. The cost of water, equipment, labour, chemicals, energy, production down time and waste treatment should be considered;

(f) the chemical supplier should be approved as per the supplier approval procedure (see Chapter 16). The chemical supplier should be able to demonstrate:

- the suitability of the chemicals for the tasks being undertaken;
- the efficiency of the chemicals especially with regard to key pathogens/microorganisms (see 14.7). The potential for spore formers surviving the disinfection process should also be considered;

[1] http://www.food.gov.uk/multimedia/pdfs/ecoli-control-cross-contam.pdf.

- a knowledge of current people's health and safety and environmental legislation and how it pertains to their products;
- an ability to provide material safety data sheets for the chemicals supplied;
- an ability to undertake the required level of operator training; and
- be able to demonstrate how the chemicals should be used in practice to deliver the required level of hygiene.

(g) the plan should take into account the areas to be cleaned, the structure and layout of the premises and the type and condition of the surfaces for floors, walls, ceilings and doors. It should also differentiate between food and hand contact services that require a disinfection process to take place and non-hand and non-food surfaces where cleaning alone may be sufficient.

(h) the plan should also take into account the equipment type, design, purpose and the type of cleaning that is required. Some equipment can be fully cleaned without dismantling while other equipment may require full dismantling. Cleaning may involve both partial cleaning and deep cleaning. Other equipment may involve a lot of pipework and a cleaning-in-place (CIP) system is used (see 14.12).

(i) the water hardness will affect the chemicals used;

(j) the type of soiling whether it is general debris, grease and/or material with a high microbial loading;

(k) the presence of services, water, steam, electricity and drainage;

(l) the supervision and training needs of the staff, as well as the mechanism for assessing the effectiveness of training and the requirements for refresher training.

(m) the records that need to be completed following cleaning; and

(n) an inventory and stock rotation system should be set up to ensure that the actual volume of chemical used complies with expected volumes. Any discrepancy between actual and expected usage must be fully investigated by the appropriate personnel.

14.9 It is important that the quality control manager has sufficient understanding and is competent in the development of cleaning protocols. It is critical that (s)he can distinguish between the different terminology and be aware of the potential issues if an incorrect type of chemical is used. All chemicals used in food environments should be food grade, non-toxic, non-corrosive and non-perfumed, and should not cause tainting.

14.10 Care should be taken to avoid mixing of cleaning agents since their chemical nature may cause them to interact and could result in a health and safety hazard.

14.11 Detergents are formulated to remove soil and dirt. Disinfection can be achieved by using:

(a) heat and/or steam as moist heat is most effective. Steam cleaners can also be used to disinfect machinery or surfaces. However the use of heat to disinfect is costly and often impractical hence chemical disinfectants are now widely used, and there is always a relationship between temperature and time; for example, equipment may be disinfected in sterilising units where equipment is immersed in water at 82°C for 30 seconds.

(b) chemical—the chemical disinfectant used will depend on a number of requirements including:

- the amount of grease and soiling;
- the effectiveness of the chemical against the type of microorganisms under consideration;
- the temperature and chemical contact time;
- the equipment and type of surfaces;
- water hardness;
- the potential for taint;
- the method of application; and
- the detergent that has been used prior to disinfection.

Disinfectants or sanitisers should meet the requirements of BS EN 1276:1997 or BS EN 13697:2001. Liquid hand wash that has disinfectant properties should conform to the requirements of the BS EN 1499:1997 standard. This information on disinfectants and sanitisers should be available on the label of the product, or may be obtained from the supplier or manufacturer. A sanitiser is a chemical that both cleans and disinfects. A sanitiser reduces the number of cleaning steps because it has both detergent and disinfectant properties and does not require an intermediary rinse. However, sanitisers are not effective on heavy soiling and can be expensive.

Cleaning in Place (CIP)   14.12   Food production systems especially in dairies or drink manufacture often have a number of bulk tanks, interconnecting pipework and pasteuriser systems. It would be impracticable to dismantle this system for cleaning and personnel entry into bulk tanks would not be safe, so the equipment is cleaned 'in place'. CIP allows equipment and pipework to be cleaned between processing runs. When designing a CIP system, the following factors are important:

- the cleaning requirement for each piece of equipment, run of pipework or tank/vessel;
- flow rate, temperatures and chemical concentrations;
- the mains services required for CIP cleaning, for example, water temperature, automatic dosing systems and waste systems;
- the time available for cleaning;
- the programme of chemicals to be used and the temperature of the cleaning chemicals; and
- the possibilities for recycling of detergents.

A schematic plan should be available to demonstrate that the CIP system has been hygienically designed and validated (see 14.7), and a log of changes and subsequent re-validation is maintained. It is very important in the development of CIP cleaning systems that:

- the automatic chemical dilution equipment is routinely monitored to ensure it is operating correctly and the chemicals are of the required dilution;

- there is adequate separation between lines being cleaned and lines filled with products, for example, double seat valves, or blanks in pipework;
- there are no dead legs in the pipework that are not cleaned;
- that in-line filters do not impede the flow rate required and that they are cleaned as part of the process and inspected at a frequency defined by risk assessment;
- that transfer pipes are not put into the system as a temporary measure and then are not cleaned because they are not part of the CIP system;
- the cleaning chemicals are non-foaming;
- the in-line pumps are of sufficient capacity to give the flow rate required;
- the temperatures are appropriate to ensure effective cleaning;
- the cleaning times are validated to ensure that there is sufficient contact time for the chemicals with surfaces;
- the levels of chemical in stock can be determined to identify if the correct amount of chemical has been used;
- the rinse water is checked to ensure that all chemical has been removed, before the system is put back into production, for example, pH monitoring;
- the location and mesh size of filters is determined as well as the monitoring activities associated with the filters;
- spray-ball or rotary cleaning head systems in tanks are checked to ensure that all areas of the tank receive a coverage of chemical and there is an adequate spray pattern; and
- all flexible hoses are cleaned as part of the CIP routine. Flexible hoses should also be capped and stored in a designated area when not in use.

CIP systems must be validated (see 14.7), monitored and verified to ensure continued effectiveness. Activities that should be undertaken include analysis of rinse water and/or first product through the lines (which is sent to waste until approved), usage data for cleaning chemicals and ATP bioluminescence testing.

Efficiency of Cleaning

14.13 The efficiency of cleaning and sanitisation should be checked and recorded at routine intervals by the quality control manager. The frequency of checks should be based on risk assessment. The standards required should be defined. Mechanisms of monitoring and verification of the effectiveness of cleaning include:

- visual inspection;
- contact microbiological swabs;
- ATP bioluminescence techniques;
- food material and product testing, where risk assessment has identified the potential for chemical residues to be present in food if cleaning procedures are not suitably followed;
- microbiological testing of part-processed and finished products; and
- microbiological and chemical checks of rinse water.

The monitoring of cleaning should be formally documented, records maintained, any trends analysed and, where required, corrective action implemented to improve cleaning and sanitisation practices. Follow-up inspections should be undertaken to ensure that the corrective action has been implemented and is effective. These records should be maintained as part of the due diligence system.

Equipment and
Storage

14.14 There must be controls in place to ensure safe and secure storage of cleaning and sanitisation chemicals when they are not in use (see 18.15). The quality control or hygiene manager, as appropriate, should develop procedures to ensure that:

- chemicals are kept in secure, closed, labelled containers and used according to manufacturers' instructions;
- chemical stores are secure, locked and bunded and away from food areas;
- stock is rotated correctly and chemicals are used within their duration marking;
- food containers are not used for storage of cleaning chemicals;
- health and safety data are available and reliable for all chemicals;
- cleaning chemicals for food processing areas and for toilet areas where these are non-food grade are stored separately;
- correct protective clothing is used; and
- spillage procedures are developed and implemented.

14.15 Appropriate cleaning equipment must be used. It should be fit for purpose and not be a source of contamination in itself. Therefore consideration should be given to using non-wooden equipment, equipment that is so coloured that a foreign object can be easily identified in the finished product, for example, brush bristles. Cleaning equipment that is used in designated areas should be colour coded to prevent its use in another area, for example, external equipment, internal machine, floor cleaning, internal factory, toilets and welfare areas. This is especially important in manufacturing units were there are designated high- and low-care areas. A colour-coded site plan may prove effective in communicating where equipment can be used. The use of mops, especially string mops, should be risk assessed to determine that they can be effectively cleaning and disinfected after use.

Cloths should be adequately controlled and be single issue, where deemed appropriate within the hygiene risk assessment. Surfaces should be allowed to air-dry, but if the task procedure requires that they must be wiped after cleaning, this must be with a single-use cloth, a sanitising wipe or blue paper towel that is appropriately disposed of to prevent contamination. If an item is found to be incomplete, for example, dustpan and plastic handle broom, then the brittle material control procedure should be implemented (see Sections 12.36–12.42). If disposable gloves are used by personnel during cleaning, they should be adequately controlled (see 5.8). The equipment should be cleaned and stored so as to minimise the potential for product contamination. Equipment should be allowed to air-dry in designated racks.

15   INFESTATION CONTROL

Principle   *The protection of food against contamination from pests is a key prerequisite within a good manufacturing practice (GMP) system. Preventive methods include staff awareness training, adequate proofing, an effective pest control programme, minimising harbourage and potential food sources and suitable control of incoming ingredients and materials that could support pest infestation. The major emphasis, however, must always be prevention.*

UK/EU Legislation   15.1   **The Prevention of Damage by Pests Act 1949** requires an occupier of any land or buildings to notify the local authority of rodent infestation. It also requires all local authorities to take steps to ensure that their district is free from rats or mice.

15.2   **The Food Safety Act 1990** made it an offence to sell food, which is unfit for human consumption or contains foreign bodies, for example, pests, parts of pests or droppings.

The EU Regulation (EC) No 852/2004 on the hygiene of foodstuffs requires that 'The layout, design, construction, siting and size of food premises are to . . . permit good food hygiene practices, including protection against contamination and, in particular, pest control'. To do this requires measures including:

- proofing of entrances and other access points;
- insect screens on open windows;
- electronic fly killers (EFKs) otherwise known as insectocutors;
- minimising harbourage especially good stock rotation of dry goods;
- regular surveys by competent pest control contractors; and
- the use of appropriate baits and chemical pest control measures.

15.3   The principles of pest control are to limit food supply, access or movement of pests, and potential harbourage. There are two types of pest control: physical and chemical methods. Physical control methods are preferable as the pest is caught, cannot further contaminate the food and there is no risk of chemical contamination of food. Physical methods also include proofing of the building. Physical methods are not always completely effective, and a combined approach with chemicals may need to be used. A major concern with chemical control is that the pest may not be killed at the point of contact, for example, ingestion, and could die later and possibly contaminate the food.

15.4   Incoming ingredients and materials should, as appropriate, be thoroughly inspected for signs of pest infestation as part of the quality control programme. All materials should be stored (see 15.10) to minimise the risk of infestation.

15.5   The most effective contribution towards infestation control is in maintaining good housekeeping standards, for example, controlling accumulations of food and paper debris, keeping gangways and passages clear, removing redundant equipment and materials from the manufacturing areas, and ensuring good stock rotation. External housekeeping is as important as internal housekeeping. External housekeeping includes:

- keeping vegetation short and not allowing vegetation to grow close to buildings;
- managing waste, ensuring all external waste areas are enclosed and they do not encourage pests nor provide harbourage;
- ensuring that leaking taps are not providing a source of water for pests to drink;
- ensuring waste building materials, redundant equipment, pallets or wooden boxes are not allowed to accumulate close to manufacturing units; and
- ensuring attention is given to other activities in the vicinity, which could attract pests, for example, adjoining farming activities, food manufacture or storage or waste handling.

15.6   All premises should either have personnel trained in pest control or employ a professional infestation control organisation for regular inspection, advice and treatment to deter and destroy infestation. An outside contractor should be accompanied on his/her visit by an appropriate staff member. It is important at the end of each visit to have a closing meeting with the contractor to discuss key points such as the current level of pest activity; contractor recommendations on proofing or general housekeeping; areas where the contractor has not been able to gain access to examine bait points; and whether any actions recommended in previous visits have been addressed appropriately by the manufacturing organisation or if further action is required and if the documentation in the pest control file is up to date.

15.7   Site inspections should be regular, and reports of each inspection should be kept on file to be available for future reference if required (see 15.16).

15.8   Care should be taken to ensure that bait spillages cannot pose a risk to any foodstuffs being manufactured or stored in the vicinity. Wherever practicable rodent baits should be based on fatty or waxy substrates. Risk assessment of individual locations should be undertaken to determine whether loose or solid bait materials will be used. The risk assessment should also assess whether only non-toxic baits should be used in internal areas. Private standards differ in their prescriptions on this point and their requirements should be determined in order to ensure compliance and thus approval. Standards, such as the British Retail Consortium (BRC) Global Standard for Food Issue 6, prescribe that only non-toxic baits can be used within production or storage areas where open product is present except when treating an active infestation (see 15.18).

15.9   No substances or application methods should be used in infestation control, which have not been approved under the **Control of Pesticides Regulations 1986**. Due attention should be given to risks of cross-contamination, for example, by spillage of grain bait, inadequate cleaning after chemical application to control pests and dispersal of tracking dust. Care should be taken to protect all materials, product, packaging, utensils and surfaces in contact with food from residual contamination by pest control substances.

15.10   Goods and equipment should always be stored at least 50 cm from adjacent walls to facilitate cleaning and inspection for rodents and insects. Drains and gutters should be fitted with screens and traps to prevent pest entry. Attention should also be paid to drains and drain pipes that could allow pest ingress.

15.11   Except where there are risks of dust explosions, all production areas should be supplied with electrical insect killing 'knock-down' devices. These should be sited in areas of minimum light intensity for best effect, but should not be sited directly above any tanks, hoppers, conveyors, processing equipment, filling machines or any open food. Pheromone traps may also be used, where appropriate.

15.12   Depending on insect type and the products being produced, pheromone traps may be used. Insect 'knock-down' devices should be fitted with suitable catch trays that should regularly be inspected and emptied when necessary (see 15.16). These devices should be left switched on at all times even after production has finished and the premises vacated; this is particularly important during the summer months. The ultraviolet (UV) light tubes on these units should be shatterproof in design and replaced at least annually. The glass tubes should be suitably controlled under the site's brittle material control procedures. The annual bulb change of the UV light should be recorded, plus any damage or breakage to the light bulb and the actions taken (see Chapter 12, Sections 12.36–12.42).

15.13   Domestic animals, for example, cats and dogs, should be excluded from manufacturing and storage areas. This is mainly accomplished by keeping all doorways and other entrances closed. Staff should never feed or otherwise encourage stray animals. Waste skips and waste storage areas should be adequately controlled to prevent attracting feral cats especially on meat processing sites.

15.14   Birds and insects must be excluded from all production and storage areas. To do this effectively, all apertures in the roof or its eaves, or the walls should be identified and either closed off or suitably screened. Drains and guttering should also be fitted with screens and traps to prevent pest access. Doors and windows similarly should be suitably protected, for example, by use of air curtains, strip curtains and netting (see 12.11).

15.15   Where there are already birds in the confines of the premises, they, and their nests, must be removed. **In the UK there are legal**

requirements under the Wildlife and Countryside Act 1981; elsewhere due regard to the requirements of the various Bird Protection Agencies must be observed.

Pest Control Programme

15.16   The pest control programme should address the following steps:

(a) *Assessment:* The pest control programme can be undertaken internally by suitably trained staff, or a contractor could be engaged to undertake either part or the entire pest control programme. If an organisation is using a pest control contractor, they must determine the suitability of the contractor and should assess criteria including the following:

- if the manufacturing organisation's customers have an approved list, and one of these approved organisations must be used;
- the proximity of the pest control contractor to the manufacturing site, resources available and the response time in the event of a problem, information such as number of employees and holiday cover, and ease of communication, for example, mobile telephone numbers and weekend cover;
- whether the pest control contractor can demonstrate technical competence; that is, in the UK, a member of a recognised association such as the British Pest Control Association (BPCA; http://www.bpca.org.uk) or National Pest Technicians Association (NPTA; http://www.npta.org.uk) and the individual operators have formally recognised training and/or qualifications and if required have been on refresher courses to keep their knowledge and skills up to date;
- the experience of the pest control contractors in the food industry sector and whether they can supply references from current clients;
- the type of contract the pest control operator provides and whether it meets internal/external requirements in terms of types of pests covered and the frequency of visits, that is, the ability of the contractor to undertake a complete pest risk assessment, undertake routinely the pest survey required and provide a clear report of their recommendations and actions required. The contract needs to address the frequency of the contractor's field biologist quality assurance surveys as well as the number of routine pest control visits; and
- evidence of the financial viability of the contractors and their current level of insurance cover with regard to product, public and employer's liability.

(b) *Protocol:* The pest control contractor should provide a pest control manual, which includes the following:

- a health and safety risk assessment for the pest control contractor employees identifying any health and safety problems. Material safety data sheets must be retained along with the Control of Substances Hazardous to Health (COSHH) assessments for the insecticides or rodenticides and other substances that the pest control contractor could use to demonstrate compliance with the **COSHH legislation** (or the equivalent outside the UK). These data should also be available for non-toxic baits if they are used. The pest control contractor in the UK must demonstrate compliance with **The Control of Pesticides Regulations 1986;**

- a contract for the pest control service, based on risk assessment, identifying the minimum number of routine inspections, emergency call-out requirements, quality assurance inspections and the types of pests controlled, that is, the scope of the pest control programme. The risk assessment should also consider external risks based on location, incoming and in-process materials and products produced, for example, stored product insects will be a factor with dry materials, and the combination of physical and chemical pest control methods that will be used. The pest risk assessment may also identify the number of field biologist visits required per annum;

- a dated bait plan, pheromone trap and/or EFK plan identifying the premises or site layout and where the baits or equipment are located. The bait points/EFK should be individually numbered and identified on the plan, and this number should correspond to the actual number on the bait box or item of equipment. The bait boxes should be robust and tamper proof, that is, the hard plastic rather than paper box design, and also be secured to prevent loss or accidental damage. In the event that a bait box is found to be missing, the quality control manager or designate must undertake an investigation to identify the whereabouts and/or potential risk to the product and action must be taken as appropriate. The bait plan should include areas of roof void as well and any baiting that has been undertaken. It should be signed off by both the appropriate member of staff from the manufacturing site and the appropriate person from the pest control contractor company. The bait plan should then be reviewed and re-signed and dated on an annual basis to demonstrate that verification activities have been undertaken and that there have been no changes. In the event of changes, the bait plan will need to be reissued and authorised by the two personnel as previously described;

- a checklist for bait points and EFK, which identifies the status of each point on each inspection for easy reference and analysis of trends. The coding is usually R = rat, M = mice or I = insects. The EFK should be individually identified. The trays should be routinely inspected and data collated including the numbers of flies, the types of flies and any trends noted. Action must be taken as appropriate;

- inspection reports should be maintained by the pest control contractor for each inspection identifying the pest control status, any evidence of pests, proofing or poor housekeeping and any recommendations. The reports should be numbered for easy cross-referencing. Any required corrective actions need to be signed off by the company to demonstrate completion. The reports should be routinely analysed by the quality control manager for trends;

- current membership certificates should be in the file for the BPCA or equivalent; and
- training certificates should be retained in the file to demonstrate the competence of the operators undertaking pest control and those undertaking quality assurance activities.

(c) *Implementation:* The pest control contractor needs to implement the agreed programme, and personnel on the site need to be vigilant to ensure that they continue to check for evidence of pests. The company should never effectively delegate responsibility for pest control to the contractor, partly because it is site personnel who are most likely to see evidence of pests or the pests themselves. Staff must be made aware that they must report evidence immediately to their line managers so that the company is actively managing its pest control system. The pest control contractor should be accompanied by a competent member of staff to ensure any issues are suitably communicated.

(d) *Evaluation:* The effectiveness of infestation controls/the pest control programme and the compliance of the pest control contractor with the requirements of the contract both need to be evaluated. An audit of the pest control procedure needs to be carried out as part of the internal audit process (see Chapter 7). The management control of the design, maintenance and proofing of buildings is critical to the pest control programme. It is not enough to have a pest control contractor visiting the site if the doors are left open or do not fit close enough to the floor, fly screens are broken or drains, roof spaces, piping and ducting allow access. The effectiveness of the premises hygiene protocols should also be assessed and its potential impact on the pest control programme. If potential food sources are not adequately controlled, then this will reduce the effectiveness of the pest control programme. Trend analysis is an important element of this evaluation.

15.17 The quality control manager and their nominated deputy must be responsible for the implementation of the pest control programme. They should be responsible for undertaking a review at regular intervals with the pest control contractor to ensure that the required level of service is maintained, for example, the number of visits complies with the contract and that the programme provides effective control, there is a hazard data sheet in the file for each chemical control used and so forth.

15.18 The initial risk assessment must determine the locations for toxic and non-toxic rodenticide. Toxic bait should not be used in areas where materials and products are subject to open storage, processing or other associated manufacturing areas unless a risk assessment has deemed that the risk level is such that the potential contamination hazard is constantly controlled to an acceptable level. If toxic baits are used as a result of an active infestation, they must be adequately controlled to prevent contamination.

16   PURCHASING

Principle

Suppliers should be selected on their ability to supply product or service that meets defined requirements. The exact nature of the supplier approval and performance monitoring will depend on the effect of the purchased product or service on the finished product. The control of incoming products and services is a key prerequisite within a good manufacturing practice (GMP) system.

General

16.1   The management team should develop the supplier quality assurance and performance monitoring procedure that defines the criteria for selection, approval, review and ongoing approval to ensure that purchased products and services meet the organisation's requirements. The extent and nature of the procedure will relate to the food products produced and the interface of materials and services purchased with the ability of the manufacturer to produce safe and legal product(s) to the required quality standards.

Supplier Quality Assurance

16.2   The quality control manager should have an advisory role in the choice of suppliers of raw materials, packaging and services. This should include determining if the supplier possesses certification to third-party assurance technical/management standards or internal company standards, certificates of analysis/conformity and/or in-house checks. (S)/he should liaise with actual or potential suppliers in agreeing the relevant specifications, including the assessment of their ability to consistently meet those specifications, and should have authority to exclude any actual or potential supplier or carrier on the grounds of inability to meet the relevant specification or of unsatisfactory performance or inadequate quality control resulting in unreliability of product or service. Assessment of the current status of the supplier's food safety management system (FSMS), hazard analysis critical control point (HACCP) plan and quality management system (QMS) should also be undertaken as relevant to the materials and services supplied. There should be continuous surveillance of supplier performance and internal review of the outputs of that surveillance at designated intervals. The intervals determined should reflect the supplier's history of conformance with company requirements. An approved supplier list should be maintained and routinely reviewed, and this activity should be defined in the supplier quality assurance and performance monitoring procedure.

16.3   The purchasing manager should have authority to place orders with suppliers on the approved suppliers list for raw material lots that conform to the relevant specifications. For those raw materials customarily sold on 'buying sample', the quality control manager should examine and report on the buying sample, and the purchasing manager should not have authority to disregard adverse findings.

Food & Drink – Good Manufacturing Practice: A Guide to its Responsible Management, Sixth Edition.
The Institute of Food Science & Technology Trust Fund.
© 2013 John Wiley & Sons, Ltd. Published 2013 by John Wiley & Sons, Ltd.

Where a delivery of a purchased raw material is found not to conform to the relevant specification or the buying sample, the purchasing manager should initiate and pursue appropriate action with the supplier as defined in the supplier quality assurance and performance monitoring procedure.

16.4   The materials and services that should be within the scope of the supplier assurance and performance monitoring procedure include:

(a) raw material and ingredient suppliers both direct suppliers, which could include agents and distributors, and if required primary producers;
(b) packaging and food contact materials suppliers including primary and secondary packaging suppliers;
(c) water suppliers;
(d) equipment suppliers;
(e) contract services such as maintenance and servicing; calibration; contract cleaning and retail tray washing services, product testing and laboratory services; transport and distribution services; training providers; contract labour and agency staff; pest control; laundry and catering services; hygiene and waste management and disposal services;
(f) subcontract manufacturing and storage operations; and
(g) processing aids.

16.5   Ingredient and direct food contact materials will require specific risk or HACCP-based assessment to determine the areas that are critical to ensure food safety, legality and quality. This assessment will include, but is not limited to, the food safety hazards addressed in Chapters 3, 4 and 12; 12.22. Exceptions to the procedures should be identified and what actions will be taken on a case-by-case basis, for example, where raw material suppliers are prescribed by retail customers, where products are purchased from cooperatives, marketing organisations or agents and where emergency procedures have meant that an alternative supplier is required at short notice to ensure continuity of supply and the timescale does not afford full compliance with the supplier approval and performance monitoring procedure. Factory trials and product development ingredients may also be deemed as exceptions; however, adequate procedures must be in place to ensure that food safety, legality and quality are fully maintained.

16.6   Supplier approval and performance monitoring should be developed on a risk basis and can include, but is not limited to, the following:

• approval of pre-supply samples including product performance standards;
• determination of the volume of material supplied and the inherent food safety hazards associated with the material;
• assessment of supplier history and their level of historic compliance with requirements;
• pre-audit of suppliers' premises to ensure they meet the required standards especially in terms of hygiene and the implementation

of prerequisite programmes (PRPs) as well as the standards defined within this guide or designated system or customer standard;

- pre-audit of supplier QMS from either documentary evidence or completion of a supplier questionnaire. In some cases, both may be necessary. The timescale for reissuing of questionnaires should be determined in the supplier approval and performance monitoring procedure and be based on risk assessment. Suppliers should be required to advise the manufacturer if there is a change to the responses they have made to the 'live' questionnaire, for example, a loss of certification, product withdrawal or recall so that the manufacturer can continue to review performance on the basis of all available facts;
- monitoring of their performance during a 'trial' period to determine the supplier's ability to deliver products and services that comply with specified requirements;
- monitoring of their performance over time in complying with the appropriate specifications;
- demonstration by means of a current certificate that the company is accredited to a recognised quality system standard, for example, BS EN ISO 22000:2005 or British Retail Consortium (BRC) Global Standard for Food Safety;
- certification to an appropriate QMS standard for distribution, transport, forwarding, shipping and air transport companies, for example, ISO 9000 or relevant BRC standard;
- certificates of analysis or conformity with each delivery;
- development of material sampling and verification procedures to check the veracity of supplier information and pre-supply samples on an ongoing basis. This could include pesticide residue analysis, heavy metal analysis and microbiological analysis. These sampling and verification activities may be undertaken by the manufacturer or supplier as formally agreed;
- review in the event of the identification of an emerging food safety hazard as a result of new scientific or technical information or a food safety incident that has been identified within the food supply chain;
- follow-up of instances of non-conformity of materials and/or service; and
- supplier audits and/or inspection checks, where practical, periodically made on suppliers based on their risk assessment.

This information is crucial to ensure that suppliers provide materials and/or services of the correct quality and that they have addressed any potential food safety or health and safety issues specific to the materials they supply. This information also provides input into the management review process ensuring that any non-conformity is reviewed (see Chapter 7). Supplier audits must be undertaken by competent individuals and an audit report documenting the audit findings and where necessary agreed corrective action should be available. Records identifying how the competency of company auditors and/or third-party auditors has been determined, validated and routinely verified should be retained by the quality manager. The completion of corrective actions by the supplier must be verified to ensure that they have been completed in a timely fashion and that they have been effective.

16.7   The supplier approval and performance monitoring procedure should address the actions to be taken in the event of product and service non-conformity including delisting procedures. A formal annual supplier review should be undertaken where each supplier's level of performance is reviewed and any actions required are determined and implemented. Questionnaires should be reviewed and updated at designated intervals to ensure that they continue to reflect supplier standards and third-party certification status. Responsibilities and timescales for completion should be developed for any resultant actions, which should be followed up at the appropriate time interval to ensure that they have been implemented and have been effective.

16.8   The frequency of testing of purchased materials and service should be determined by the quality control manager. Materials should be inspected on receipt to ensure compliance with specifications and that traceability has been maintained from the original source. On receipt certificates of analysis, certificates of conformity and pre-acceptance testing (positive release testing) may be required depending on the initial risk assessment undertaken. Packaging that does not undergo quality checks on receipt should be monitored as part of the production quality control checks. Material and supplier risk assessments must be reviewed on a minimum of an annual basis and when non-conformity is identified. Appropriate action must be undertaken to ensure that any changes to procedures and specifications are validated and implemented, and are effective.

Use of Outside Services

16.9   As previously mentioned in 16.4, the supplier approval and performance monitoring procedure also needs to address the range of outside services that could be purchased under contract. These include:

- building repair and maintenance;
- laundering;
- waste disposal;
- specialist engineering maintenance;
- warehousing;
- computer services; and
- advisory services such as business consultants, technical consultants, marketing consultants and legal consultants.

16.10   The contract giver should fully satisfy himself that a contract acceptor has the appropriate knowledge, skill, facilities, equipment, staff and so on to be able to provide the service in such a way as to contribute to the objectives of GMP. Except in the case of purely advisory service, if the work is carried out away from the contract giver's premises, the contract giver should have and should exercise the right to visit and inspect any location where the work is performed.

16.11　A contract acceptor should not pass on to a third party any of the work commissioned by the contract giver except with the prior knowledge and consent of the latter.

16.12　In the absence of specific authorisation to the contrary by the contract giver, the contract acceptor should treat as confidential any direct or incidental knowledge gained about the contract giver's business.

16.13　Arrangements between a contract giver and a contract acceptor should be in writing and should include adequate indication of the agreed respective responsibilities of the parties, and of any special hazards of which either party is aware, and record the authority for the contract acceptor to conduct the work.

17   PACKAGING MATERIALS

Principle

Packaging can be described as the materials used for the protection, preservation and presentation of food products. Product packaging should comply with relevant legislation and conform to agreed specifications. It should be stored in a designated area and in such a way as to minimise contamination or damage. Packaging procedures are a key prerequisite to ensure good manufacturing practice (GMP).

General

17.1   Principles of GMP with regard to packaging have been reviewed in Chapter 6 (6.24–6.35).

Storage

17.2   Packaging materials should be stored in a designated area separate from raw materials and finished products. If open containers are not stored in external packaging, then where possible they should be inverted to minimise the risk of product contamination. Packaging liners should be coloured to minimise the risk of product contamination. Where possible packing should be shrink-wrapped to provide security on the pallet and minimise any potential foreign body contamination.

17.3   Packaging materials consist of:

- primary or sales packaging;
- secondary or grouped packaging:
- tertiary or transport packaging; and
- transit packaging.

The main materials that are used for primary, secondary, tertiary and transit packaging include paper, glass, aluminium, steel, plastic and wood. Packaging materials should be suitable for the food they contain and should be inert during packing, processing, storage and distribution and in preparation by the consumer for consumption, for example, microwaving or cooking in the pack. Consideration should also be given to the pH in the case of low-acid and high-acid foods and fat content of the food and whether this will impact on packaging design and suitability. Packaging specifications should be developed and approved for all packaging materials. The specifications may be provided by the customer especially where own-label branded products are being supplied or originated by the manufacturer. Packaging specifications should contain the following information as appropriate:

- supplier name and address and manufacturing site name and address if both are different;
- material composition;
- packaging dimensions including thickness and gauge;

Food & Drink – Good Manufacturing Practice: A Guide to its Responsible Management, Sixth Edition.
The Institute of Food Science & Technology Trust Fund.
© 2013 John Wiley & Sons, Ltd. Published 2013 by John Wiley & Sons, Ltd.

- specific storage and handling requirements, for example, temperature and humidity, single or double stacking;
- material suitability and confirmation of adequate migration testing results and compliance with legislation (see 12.22);
- information on adhesives used if appropriate (see 12.22);
- artwork details including size of lettering, bar code details, marketing and farm assurance logos and colour standards; and
- promotional information if there is both standard and promotional packaging utilised for the same product.

17.4   Some transit packaging such as field trays, field bins or retail plastic trays are reusable/returnable. These trays are used, for example, as a display outer for fresh products including fresh produce and meat. These trays are sometimes referenced as returnable transit packaging (RTP). While all foods are within primary packaging, it is important that these trays are of a consistently high level of hygiene and cleanliness. It is also important to ensure that all previous tray/box labels are removed before use to ensure that identity and traceability is maintained. Risk assessment should be undertaken with loose food, for example, fruits and vegetables that are packed in a tray liner to determine whether the hygiene of the trays could detrimentally affect the food safety and quality of the product. If this is deemed a risk, then suitable controls need to be put in place and verification activities undertaken to determine their continued effectiveness.

17.5   Tray washing between retail outlets and manufacturing plants is often undertaken by third-party companies. These companies should be subject to the supplier approval and performance monitoring procedure (see Chapter 16). The quality control manager should develop procedures to ensure that retail tray monitoring procedures are implemented to ensure that hygiene levels are maintained and labels have been removed and that appropriate action is taken in the event of non-compliance.

17.6   Particular attention should be paid to the temperature of the tray wash tank (ideally 75–80°C) and the rinse tank (ideally 80–85°C). The Fresh Produce Consortium Guide Hygiene Procedures – A Code of Practice for Returnable Transit Packaging at Traywash Units (March 1998) suggests that, based on a maximum number of 1200 trays per hour, trays are exposed to temperatures for the following times:

- Wash tank   30 seconds
- Detergent rinse   7 seconds
- Final rinse   13 seconds

These temperatures and times should be routinely monitored.

17.7   Retail trays, where possible, should not be stored outside. They should be shrink-wrapped to minimise contamination.

17.8   Packaging should be stored according to manufacturer's instructions to minimise damage and possible effects on packaging integrity. For example, reels of laminate packaging need to be stored at the correct conditions and in a way to minimise damage so that the laminate remains intact and does not allow air ingress.

17.9   Heat-preserved products require the maintenance of packaging integrity to ensure control of product safety and quality characteristics. The quality control manager should develop procedures to monitor packaging integrity and packaging seal quality where this is required, for example, canning, aseptic packing, vacuum packing and gas flushing operations. This will include the need to develop appropriate sampling plans. Protocols for accelerated product durability testing should be developed where appropriate.

17.10   Packaging should be traceable on a batch basis, and the quality control manager or designate should develop a packaging traceability procedure to ensure that control is maintained. Food safety hazards associated with packaging have been discussed (see 5.2, 5.14–5.18 and 12.22). Packaging suppliers should be monitored and approved as per the supplier approval and performance monitoring procedure (see Chapter 16).

17.11   The **EC Directive on Packaging Waste (94/62/EC)** aims to reduce the volume of packaging waste going to landfill sites by setting targets for the recovery and recycling of packaging waste. The UK legislation that implements the **EC Directive is the Producer Responsibility Obligations (Packaging Waste) Regulations 1997** (see 22.3).

18   INTERNAL STORAGE

Principle

Storage areas should be so designed that they are fit for purpose, and the layout and materials used allow for appropriate and effective cleaning. Storage procedures should be in place to prevent damage or deterioration of both premises and the materials contained therein. Effective storage procedures minimise the risk of contamination and are a key prerequisite within a good manufacturing practice (GMP) system. Raw materials, packaging materials, part-processed and finished products, cleaning chemicals, personal items, equipment and machinery spares should be stored, where possible, in segregated, separate storage areas to minimise the risk of cross-contamination.

General

18.1   Access to material and product storage areas should be restricted to personnel working in those areas and other authorised persons. Consideration should be given to site security and entry controls into designated storage areas (see 5.21, 11.21s, 12.2, 19.23, 23.25 and 28.5–28.9). Designated separate lockers may need to be provided for both internal and external workwear as well as personal belongings that must not be taken into the production and storage areas (see 11.21f).

18.2   Packaging, food ingredients and products, equipment and other items should all be stored in separate, designated storage areas. Materials and products should be stored under the conditions specified in their respective specifications. Particular attention should be paid to the avoidance of allergenic or microbiological cross-contamination and tainting. Where special conditions are required, they should be regularly checked for compliance.

18.3   Materials and products should be stored in such a way that cleaning, the use of pest control materials without risk of contamination, inspection and sampling, retention of delivery identity or batch identity, and effective stock rotation can be easily carried out (see 15.10 and 18.6).

18.4   There should be effective protection of equipment spares, materials and products from contamination.

18.5   Storage areas should be so designed as to minimise the risk of contamination of stored items (see 6.52). Effective cleaning of storage premises and equipment must be carried out at the designated frequency and using the methods and materials specified in documented cleaning schedules and instructions (see Chapter 14). The effectiveness of cleaning should be verified and as a result any appropriate action taken.

Food & Drink – Good Manufacturing Practice: A Guide to its Responsible Management, Sixth Edition.
The Institute of Food Science & Technology Trust Fund.
© 2013 John Wiley & Sons, Ltd. Published 2013 by John Wiley & Sons, Ltd.

18.6   When developing storage procedures, the following should be taken into consideration:

(a) lighting, temperature and humidity control and ventilation should be adequate for the purpose. Storage areas should be designed and managed to minimise condensation (see 12.7 and 12.10), and any condensate pipes, for example, from refrigeration units should be designed to flow directly above a drain and not be allowed to drip onto the product, materials, packaging, equipment and personnel. This should be verified during premises audits. The condensate pipework should be designed to prevent airflow back into the pipework and consideration should be given to the need to sanitise the condensate produced. The pooling of water around storage areas especially where there is vehicular access should be minimised;

(b) storage areas should be kept clean and tidy to minimise harbourage or food sources for pests (see Chapter 15);

(c) damage to stored materials should be minimised and all spillages should be cleared away promptly;

(d) storage areas should have adequate proofing to prevent pest ingress and external doors should not be left open;

(e) all materials within storage areas should be protected from excess heat and light, water penetration and accumulation of foreign matter;

(f) stock rotation should be undertaken, and all items should be marked with their identification so that traceability is maintained;

(g) items stored on pallets should neither be touching the walls nor blocking the main doors or passageway. If racking is used, the layout should be designed so that the racking is far enough away from the wall to prevent pallets and palletised product being damaged or touching the wall. There should be designated vehicular and pedestrian access in storage areas with racking so that product can be safely inspected during storage. Space should be left in all gangways for product inspection to take place;

(h) any suspect stock should be segregated, ideally in an area designated for that purpose (Chapter 21); and

(i) product should not be double stacked where this should prove a potential contamination risk or could affect the integrity of packaging.

Procedures should be developed on the basis of the points raised above, then implemented and verified to ensure that they are effective and understood and followed consistently by those staff working in storage areas. This is especially important in temperature-controlled storage areas such as chilled or frozen stores and where allergens are stored on-site. Furthermore, staff should be trained to understand the need to prevent material and product damage during storage and why the maintenance of product integrity and product quality is important. Routine audits, or other appropriate verification activities, must be undertaken and recorded to demonstrate that damage of product and materials during storage is minimised. Appropriate corrective action should be implemented as necessary and should be followed up to ensure that it remains effective.

18.7 The quality control manager or designate is responsible for developing and implementing appropriate monitoring procedures for temperature-controlled storage areas to ensure that the storage area is capable of maintaining the appropriate temperature profile during work activities. The results of monitoring should be formally recorded and any appropriate corrective action taken where required (see Chapters 33 and 34).

18.8 Products that have been recalled or returned, and batches that have been rejected for reworking or recovery of materials or disposal, should be so marked and physically segregated, preferably in an entirely separate storage facility (see Chapter 21).

18.9 Material deliveries and product batches temporarily quarantined pending the results of testing, should be so marked, suitably segregated, and effective organisational measures implemented to safeguard against unauthorised or accidental use of those materials or despatch of those products.

18.10 If a batch of finished product has to be temporarily stored unlabelled, to be labelled at a later date, the greatest possible care should be exercised in maintaining its exact identity and ensuring correct durability indication when labelled (see Chapter 10).

18.11 Storage areas should be regularly inspected for cleanliness and good housekeeping, and for batches of products that have exceeded their shelf life or, in the case of date-marked products, leave insufficient time for retail display. These inspections should be formally recorded, and the records should include in the event of non-compliance any required corrective action and demonstrate that corrective action has been followed up to verify it has been effective.

18.12 Risk analysis should be undertaken by the quality control manager or designate to ensure that cross-contamination, including where relevant airborne taint, is prevented. This is especially so at times of peak production where storage space may be limited. The Food Standards Agency (FSA) publication *E. coli* O157—Control of cross-contamination: Guidance for food business operators and enforcement authorities (2011) stresses the importance of managing storage not only of ingredients and finished products, but also of packaging. It also highlights the importance of ensuring that physical separation of materials is effective especially if high-risk, ready-to-eat foods are manufactured/produced.

18.13 Pallet labelling should be undertaken so that it is clearly visible during storage. Attention should be paid to the effective adherence of pallet labels especially in storage conditions that could affect the ability of the labels to remain intact on the pallet. The number of pallet tickets applied (and on which faces of the pallet) needs to consider the visibility of pallet tickets especially when product is placed in racking. Consideration should also be given to maintaining traceability on pallets when the outer wrapping is removed and/or where pallet tickets may have been adhered to boxes at the top of the pallet that will be used first. Traceability must be maintained at all times.

Storage of Chemicals, 18.14 The stock controller or designate should be responsible for the taking
Lubricants and Oils into stock of all chemicals. (S)/he should also be responsible for ensuring that these items are as per the delivery instructions and are held, in storage, in their original packaging until required by particular personnel for cleaning or maintenance activities. All materials must be stored in sealable and labelled containers and handled and transported in a safe and responsible manner. Stores should be sound, secure, bunded (or alternatively the materials can be stored on bunded pallets), well ventilated, frost proof, have ease of access and have sufficient light to enable the operator to read the product label. Appropriate warning signs should be placed on access doors according to the inherent characteristics of the chemicals, for example, flammable and corrosive. All materials should be stored off the floor, if not on shelving then on pallets. Shelving should be made from non-absorbent material and powders should be stored on shelves above liquids. When materials are on shelves then the store must be bunded. Stores should be able to retain spillages, and emergency procedures should be in place to deal with accidental spillages. Protective clothing must be worn where applicable. Appropriate personal protective equipment (PPE) should be supplied for all operations involving chemicals, but it should not be stored in the storage area. PPE should be stored in a designated clean, dry, well-ventilated and secure locker. The minimum requirements for PPE are detailed on the chemical product label. Any additional requirements should be identified during the Control of Substances Hazardous to Health (COSHH) assessment. Protective clothing must be personal to the individual, suitable for its intended purpose, in sound condition and the correct fit for the wearer. PPE should be cleaned, maintained, stored and disposed of according to manufacturer's recommendations and statutory requirements. Empty containers must be stored as per legal requirements and must be suitably disposed of.

19 CRISIS MANAGEMENT, COMPLAINTS AND PRODUCT RECALL

Principle

The full significance of a food safety or quality complaint may only be appreciated by certain responsible persons, and then possibly only with the knowledge of other related complaints. A procedure must therefore be provided for the appropriate channelling of all complaint reports and the analysis of complaint data. A product defect coming to the manufacturer's attention, whether through a complaint or otherwise, may lead to the need for a product withdrawal from the retail distribution system, or a public product recall also involving return of products by members of the public. There should be a predetermined written plan, clearly understood by all concerned. It should address the withdrawal or recall of a product, or a known batch or batches of product known or suspected to be hazardous or otherwise unfit, or the withdrawal or recall of wholesome but sub-standard product that the manufacturer wishes to withdraw or recall. A crisis management procedure and a crisis management team should be established.

Complaints

19.1 The system for dealing with complaints should follow documented procedures that indicate the responsible person, and their deputy, through whom the complaints must be channelled.

19.2 If the responsible person is not the quality control manager, the latter should be fully informed and closely consulted. The responsible person should have the appropriate knowledge and experience, and the necessary authority to decide the action to be taken.

19.3 Where possible, product food safety, legality and quality complaints should be thoroughly investigated by the quality control manager, with the cooperation of all relevant personnel, and a report prepared as a basis for action and for the records. This report could be analysed as part of the management review process (see 7.2). This process should be robust enough to differentiate between one-off incidents and an ongoing trend.

19.4 Corrective action should include responding to the complainant, and must include responding to any enforcement authority involved. Where the complaint is justified, steps to remove or overcome the cause and thus prevent recurrence should be taken; and the defective material that the complaint sample might represent should be dealt with including possibly a product withdrawal or recall. Root cause analysis should be undertaken to identify the factors that have caused the incident so that appropriate preventive and corrective action is implemented and is effective (see 20.10–20.12).

Food & Drink – Good Manufacturing Practice: A Guide to its Responsible Management, Sixth Edition.
The Institute of Food Science & Technology Trust Fund.
© 2013 John Wiley & Sons, Ltd. Published 2013 by John Wiley & Sons, Ltd.

19.5 Complaint reports should be regularly analysed, summarised and reviewed for any trends or indication of a need for a product recall or of any specific problem requiring attention. It is strongly recommended that appropriate summaries that include comparative data are developed and that they are regularly distributed to directors and senior management and subject to formal management review (see Chapter 7). Trend analysis should relate the number of complaints to volume produced to give a better reflection of performance such as complaints per 100,000 units or complaints per million units sold/produced. Whether the complaints are related to products sold or produced will depend on the seasonal variation of production and also the products' indication of durability. Fluctuations in production levels/purchasing levels could, if not taken into consideration, potentially skew the trend analysis. Measurable indicators such as complaints per million units sold, percentage compliance with agreed service level or rating of types of complaint should be developed and form part of the management review process.

Product Withdrawal and Recall

19.6 The type of occurrence that constitutes a product incident should be determined. As there are so many types of incident that could occur, it is not possible in this publication to give a definitive list or be specific on the actions to be taken in each instance. The type of product incident likely to occur will be specific to the product and its ingredients, the process employed in preparing, storing and manufacturing the product, the frequency of reworking or regrading activities, which in turn will reflect ongoing supplier performance at meeting material specifications, and the overall effectiveness of the quality management system (QMS) and food safety management system (FSMS). An incident could be contamination due to an allergenic, biological, chemical or physical hazard whether accidental or malicious; human contamination such as blood or other material; adulteration; and/or any other incident that could affect the quality, safety or legality of the product such as incorrect labelling or incorrect use of ingredients.

19.7 A responsible person, with appropriate named deputies, should be nominated to initiate and coordinate all withdrawal and recall activities, to liaise with retailers, and to be the point of any contact with (in the UK) the Local Authority, Port Health Authority, Food Standards Agency (FSA), Department of Environment, Food and Rural Affairs (DEFRA), Department of Health and/or the Health Protection Agency on withdrawal and recall matters (or equivalent elsewhere).

19.8 The design of manufacturing records systems and distribution records systems, and the marking of outer cartons and of individual packs should be such as to facilitate effective withdrawal or recall if necessary. A good system of lot or batch marking will pinpoint the suspect material and help avoid excessive recall. Lot marking of most food products is a requirement of the **UK Food (Lot Marking) Regulations 1996**, implementing the **EU Directive 89/396/EEC**, as amended by **Council Directives 91/238/EC and 92/11/EC** (see Chapter 10).

19.9   There should be written withdrawal and product recall procedures, and they should be capable of being put into operation at short notice, at any time, inside or outside working hours. To ensure this is possible, an out-of-hours contact list should be maintained and include contact details for staff, contractors, suppliers, customers, emergency services, specialist laboratories, legal advisers, certification bodies and government and enforcement regulatory bodies. Procedures should also include a communication plan, which details the communication levels both internal and external to the organisation as well as media communication that would be used in a full product recall.

19.10 The procedures should be shown to be practicable and operable within a reasonable time by carrying out suitable testing of the procedures. The product withdrawal/recall and management of incidents system must be tested at least annually to determine the effectiveness of the system both for traceability from the finished product to raw materials and packaging materials used and from a batch of raw material to the individual batches of products that it was used to make.

19.11 It is important to test the product withdrawal/recall procedure for a range of incidents. For example, a test could be undertaken for either glass contamination, metal contamination or an instance of contamination at the supplier. Each type of incident will require a different audit trail, and only by rotating the type of incident will the organisation be able to demonstrate the true effectiveness of the product recall and withdrawal procedure. If the product recall test is only undertaken to test the ability to trace a production code to retail depot or from depot to supplier, it is in effect only a traceability test. Thus, in the event of a glass contamination incident, the ability of the organisation to collate the data and records required will never be tested and could be fundamentally flawed. This will then only become apparent in a 'live' situation. This test should be documented and records retained. Corrective action identified as a result of non-compliance during a test must be verified to ensure that it has been implemented and has been effective.

19.12 The length of time for a product withdrawal/recall test between initiation and completion should also be recorded to ensure that a potential withdrawal/recall would be carried out in an appropriate timescale.

19.13 The procedures and documentation should be reviewed regularly to check whether there is a need for revision in the light of changes in either circumstances or who is deemed to be the responsible person. This could include changes to telephone numbers, contact details or specific customer procedures such as the time limit for notification.

19.14   Product withdrawals or recalls may arise in a variety of circumstances, which however fall into three main categories:

(a) where the food enforcement authorities become aware of a
hazard or suspected hazard, and information and cooperation
from the manufacturer or importer is necessitated;
(b) where the manufacturer, importer, distributor, retailer or caterer
becomes aware of a hazard or suspected hazard;
(c) where there is no hazard or suspected hazard involved, but there
is some circumstance (e.g. substandard quality, mislabelling) that
has come to light and that prompts the manufacturer, importer
or retailer to decide to withdraw or recall the affected product.

In case (c), the company will itself have to organise the withdrawal
or recall operation. In cases (a) and (b), consideration may be given
to issuing a public food hazard warning. Generally this would be
done in consultation among the manufacturer or importer, the distributor or retailer, and any relevant enforcement authority interest.
The UK FSA issues Allergy Alerts and Food Alerts[1] based on recalls
from the retail stage of the supply chain. The FSA issues a 'Product
Withdrawal Information Notice' or a 'Product Recall Information
Notice' to let the general public, the food supply chain and Local
Authorities know about problems associated with food. A Food Alert
for Action is issued where formal intervention by enforcement
authorities is required. Normally any arrangements for recall would
be discussed so that the most appropriate methods could be effected
or endorsed by the relevant authorities, and would also take into
account any requirements for or arising from the information indicated below.

19.15   Although a defect or a suspected defect leading to withdrawal or
recall may have come to light in respect of a particular batch or
batches or a particular period of production, urgent consideration
should be given to whether other batches or periods may have also
been affected (e.g. through use of a faulty material, or a plant or
processing fault), and whether these should also be included in the
withdrawal/recall.

19.16   The procedures should lay down precise methods for notifying and
implementing a withdrawal or recall from all distributive channels,
retailers and of goods in transit, that is, wherever the affected product
might be. It should also include a procedure to prevent any further
distribution of affected goods. The recall procedure should also
provide for method of public notification.

19.17   Notification of withdrawal/recall should include the following
information:

(a) name, pack size and adequate description of the product;
(b) identifying marks of the batch(es) concerned;

[1]http://www.food.gov.uk/subscribe/.

(c) the nature of the defect;

(d) action required, with an indication of the degree of urgency involved.

19.18   Recalled or withdrawn material should be quarantined, pending decision as to appropriate treatment or disposal.

*Emergency
Procedure*

19.19   Regrettably the possibility of real or threatened hazards arising from the actions of second or third parties must be faced, for example, deliberate contamination or poisoning of product or ingredient by terrorists, extremists or otherwise misguided persons. Although some of the additional action that might be taken in such circumstances could be considered outside the scope of this Guide, it is included because those concerned in the manufacturing operation would very probably become involved.

19.20   The first intimation of a problem in this area could come from a whole variety of sources, for example, consumer complaint, from a retailer, the media, the police, the enforcement authorities, employees, or by telephone, email or texting, post or personal contact with any company location or any employee at any time.

19.21   It is therefore essential that any personnel engaged in manufacture should be aware of company procedures to be followed in dealing with such threats both within and outside of normal working hours, and that suitable arrangements exist for calling in key personnel out of hours in such an emergency. The extent to which any such emergency procedures may override normal lines of management should be explicitly stated. These procedures should be formally documented.

19.22   Faced with an emergency situation, the withdrawal/recall procedures described above will apply, while the expertise of those involved in quality control and other relevant functions should be put at the disposal of the crisis management team responsible for handling the emergency.

19.23   The possibility of such sabotage, vandalism, terrorism and even site invasion may indicate a need for particular security precautions in vulnerable areas, for example, entrance security, closed-circuit television (CCTV) security, code pads to open external doors to manufacturing areas, locked rooms and the use of tamper-evident or other type of security seals (see Chapter 5, Section 5.21).

19.24   Any emergency or recall situation is likely to involve retailers, wholesalers or caterers, and a smooth and effective interface with their procedures should be achieved as early as possible during the crisis.

19.25   The potential actions of current or previous employees need to be addressed by the manufacturer including the need for pre-employment screening and employment termination interviews.

Business Continuity Management

19.26 An important element of good manufacturing practice (GMP) is maintaining the continuity of supply in the event that an incident occurs that disrupts 'normal' manufacturing activity. The business operator should have developed written continuity plan(s) to identify the impact of specific incidents and what actions would be taken in the event that they occurred. Malicious damage and sabotage has been previously addressed in this chapter but the following should also be considered:

- natural disaster, environmental release, for example, refrigerant, fire and flood;
- disruption of services including water, energy, transport and logistics; communications, for example, Internet and telephone; staff availability including a food safety outbreak; and a failure in temperature control equipment.

20   CORRECTIVE ACTION

Principle

A procedure must be provided for appropriate action to be taken in the event of product or service non-conformity that could impact on product safety, legality or other specified quality characteristics. This procedure needs to provide for the actions to be taken on identification of non-conformance, determining the extent and the implications of that non-conformance and the appropriate actions to be taken.

General

20.1   Major non-conformance is defined as a food safety management system (FSMS) or quality management system (QMS) failure or fault, which could give rise to a food safety hazard, or a major product defect. However, if minor non-conformance persists or if a trend is noted in a series of minor non-conformance, this sequence of events could be deemed major.

20.2   A procedure should be developed and implemented by the quality control manager to ensure that, once identified, non-conforming items are clearly identified, segregated and/or otherwise controlled so that unauthorised or inadvertent use is prevented and disposal is formally agreed and recorded (see Chapters 19 and 21). The procedure should also ensure that all events of non-conformity are fully and adequately recorded.

20.3   Routine quality control inspection or monitoring incorporated within the FSMS may identify instances of non-conformance. Quality control procedures must identify the actions to be taken in the instance that non-conformance of product or service is identified.

20.4   Corrective action procedures should be established not only for the manufacturing process, but also during storage and distribution including the transport of raw, part-processed and finished products.

20.5   All corrective action should be documented on the appropriate records that identify:

- the date the form is completed;
- details of the actual nature of the non-conformity;
- product details including traceability information so that the batch(es) involved can be identified;
- how much product is affected;
- location of the product, for example, on-site or through the distribution chain;
- the root cause of the non-conformity;
- the required action to be taken including the need for product recall or product withdrawal;

Food & Drink – Good Manufacturing Practice: A Guide to its Responsible Management, Sixth Edition.
The Institute of Food Science & Technology Trust Fund.
© 2013 John Wiley & Sons, Ltd. Published 2013 by John Wiley & Sons, Ltd.

- the timescale for the required action;
- who is responsible for undertaking that action;
- that all parties have accepted the non-conformity and agreed the appropriate corrective action;
- who is accountable for the decision on the corrective action necessary; and
- who is responsible for ensuring corrective action has been undertaken and is effective.

20.6   Internal audits may identify evidence of non-conformity and necessitate the requirement for corrective action. If there are a number of corrective actions, the auditor should develop a corrective action plan (see 7.9). Corrective action plans should address the following:

- ensuring all raw materials, ingredients, processing aids, packaging and part-processed and finished products are safe, legal and comply with specifications;
- minimising any impact on the customers and ensuring the non-conformity does not occur again;
- the importance of the non-conformity—whether the non-conformity is minor—a one-off incident, a repeated minor incident or a major system failure; and
- the resources required to implement corrective action in terms of financial resource, training, personnel and the development of documentation and follow-up activities to ensure it has been effective;

20.7   Corrective action plans need to be reviewed by the quality control manager or designate to ensure that the required actions are being undertaken by the designated date and the actions are effective. The corrective action plan needs to take into account both the immediate corrective action to ensure that the non-conforming product is brought back under control and the longer-term corrective action that addresses the root cause of the non-conformance. Root cause analysis can take many forms and follow specific mechanisms, but essentially is a structured management approach that identifies the factors that resulted in the non-conformance in order to determine the most appropriate corrective or preventive action. The factors that could be considered include the actual nature of the non-conformance, the magnitude (major or minor) and the consequences of the problem in order to identify the actions, conditions or behaviours that need to be changed to prevent reoccurrence and/or other similar problems from occurring. Conditions can include processing parameters, procedures, protocols or standards implemented in the manufacturing operation.

20.8   Areas of non-conformity and the status of corrective actions as well as the effectiveness of previous corrective actions provide input to the management review process (see Chapter 7).

20.9   Corrective action must be followed up at an appropriate interval to ensure that the corrective action has been implemented and has been effective. Verification activities to assess effectiveness of corrective action include:

- review of monitoring results – giving feedback on the effectiveness of the corrective action;
- product testing, for example, shelf-life testing and microbiological testing; and
- follow-up audits to ensure that the non-conformity identified is now under control.

Customer Complaints/Depot Rejections

20.10 Corrective action arising as a result of customer or consumer complaints and/or customer depot rejections needs to be formally controlled. The quality control manager should develop procedures that define how all quality aspects of distribution activities will be coordinated (see Chapter 19).

20.11 The quality control manager, or designate, should make a formal assessment of the product on its return to determine the correct disposal. The quality control manager should develop an appropriate documented procedure with associated forms to control this process including the responsibilities of key individuals in the formal decision making. Routine analysis of the records should be undertaken to determine trends and any additional corrective action that may be required.

20.12 If product is reworked or regraded and then redistributed, records must be maintained to ensure traceability to the original batch details and associated records (see Chapter 21).

21   REWORKING PRODUCT

Principle

Material may be recovered, reworked or reprocessed by an appropriate and authorised method, provided that the material is safe and suitable for such treatment, and traceability of original raw materials and part-processed product is maintained so that the resulting product complies with the relevant specification and that the related documentation accurately records what has occurred.

General

21.1   Consideration should be given to how reworked material is assessed to ensure that a reworked material is not in itself a potential contaminant. Examples are:

- material from products containing ingredients or additives such as preservatives and colourants not present in the intended recipient product in which the material will be reworked;
- material from products being reworded that do not comply with the specific production method (e.g. halal, kosher, organic, vegan, vegetarian) of the intended recipient product;
- material being reworked not complying with the compositional profile of the intended recipient product (e.g. health products such as diabetic, low carbohydrate or low fat, or products suitable for a low-calorie diet);
- material or intended recipient product of a different specified provenance, country of origin, assurance status or designated geographic status so that identity preservation is lost; and
- material from products containing food allergens being reworked into the intended recipient product where allergens are absent (see 4.4).

All reworking activities must be undertaken in line with strict procedures to ensure that the resulting finished product is safe and complies with relevant legal and quality criteria.

21.2   As there are so many different circumstances that can arise with different kinds of food products and processes, it is not possible to be specific here about each of them. The matters referred to here, however, may be classified under three main groups, namely systematic, 'semi-systematic' and 'occasional'. However, as defined in 21.1, in all circumstances, appropriate precautions must be taken to avoid microbiological contamination, introduction of undeclared ingredients, cross-contamination with allergens or the introduction of foreign matter, and to avoid the loss of traceability and provenance.

21.3   The possible carrying forward of perishable material left over from the previous day should be subject to formal risk assessment by the quality control manager. Where a quantitatively known product

Food & Drink – Good Manufacturing Practice: A Guide to its Responsible Management, Sixth Edition.
The Institute of Food Science & Technology Trust Fund.
© 2013 John Wiley & Sons, Ltd. Published 2013 by John Wiley & Sons, Ltd.

residue from the previous production is systematically utilised as one of the starting materials for the same or another product (e.g. dough trimmings in biscuit manufacture), that should be written into the master manufacturing instructions, and the rate or conditions of use there specified should not be departed from other than through the established procedure for varying master manufacturing instructions.

21.4 'Semi-systematic' applies to instances where a variable quantity of intrinsically satisfactory but extrinsically unacceptable product occurs and can be reused (e.g. misshapen or short-weight moulded chocolate bars), or to instances where a usable starting material can be extracted from wholesome but defective product (e.g. recovery of sugar as a syrup from misshapen or erroneously formulated sugar confectionery). In such circumstances, provision for such recovery should be made in the master manufacturing instructions, specifying a maximum limit to the rate of incorporation.

21.5 'Occasional' instances are any instances other than referred to in 21.3 or 21.4. They should in all cases be subject to risk assessment by the quality control manager before any decision as to disposition. The four main categories are as follows:

(a) batches of intermediate or bulk product that have been quarantined as sub-standard. In some instances (e.g. insufficient salt in a batch of soup or sauce), simple adjustment by addition of the calculated amount of deficient ingredient will suffice; in the case of an excessive amount of an ingredient, rectification may be possible by making a special batch with a calculated deficiency of the ingredient in question and blending the two batches. An alternative in the two cases mentioned, and the only possible use where other types of defect are involved, would be to rework a certain amount of the defective material into a number of succeeding batches. The amount per batch should be determined by experience or a trial where necessary, and agreed by the quality control manager, and should be the subject of a written instruction. Particular regard should be paid to any microbiological or other problems that might arise through holding the material in question for the length of time involved and how traceability can be maintained;

(b) packed finished product that has been quarantined as sub-standard. Where the defect is merely soiled, scuffed or badly applied labels, incorrect labelling or stickering, relabelling is appropriate. In instances where the product itself is sub-standard, it is rare that in-pack reprocessing would be either possible or appropriate (except in the case of certain canned products where safe heat processing has resulted in products the texture of which could be improved by further heat processing). In general, the utilisation of quarantined sub-standard packed goods (except in circumstances dealt with in (d) below) requires the emptying of the product from the packages. Its utilisation would then be subject to the considerations indicated in (a);

(c) packed finished product recalled or otherwise returned from distribution. Similar considerations apply to those referred to in (b), with the addition, however, that as the packs concerned have been outside the control of the manufacturer for a period, they should be assessed even more critically and in the light of information as to their age, history and condition;

(d) in the special case of a group of finished product packages that has failed the reference test for filled weight or volume (either on average quantity or on an excessive number of non-standard packages), various ways of rectification are possible. In the case of failure only because average quantity is too low and no package is below the legally acceptable minimum, rectification may be effected by blending the group with another group specially prepared with the average as far above the target as the defective group's average is below it. Where the failure is only through excess non-standard packages, the group may be rectified by sorting to remove all or most of the non-standard packages. Another possibility is relabelling with a lower nominal quantity. Finally, if it is feasible to open and reseal containers (where this can be undertaken without compromising food safety), topping up with material from the same batch may be used to increase average quantity or decrease the number of non-standard packages. **The reader is referred to 'Code of Practical Guidance for Packers and Importers; Weights and Measures Act 1979' (HMSO) for detailed guidance in this topic.**

Rejection 21.6 Inevitably, rejection of materials described above will be necessary from time to time, and proper means of disposal should be considered and agreed with the quality control manager, the production manager and any other interested parties such as the purchasing manager or sales/marketing departments. In determining disposal, due regard should be paid to the needs of securing cost recovery, protecting the company or brand name, protecting the public **and complying with appropriate legislative or local authority requirements**. For example, disposal subject to appropriate safeguards might range from sale to a third party for relabelling or packing as a lower-quality product, to staining and selling for including in animal feed or to seeking local authority condemnation and disposal at their hands. **Consideration must be given to compliance with all current European Union (EU) or UK environmental and food and feed safety legislation including Regulation (EC) No 1774/2002 (enacted in the UK by the Animal By-Products Regulations 2003).**

Relabelling 21.7 In any relabelling of packs, any identifying marks carried by the original labels should be carried by the new labels, and where the pack carries a durability indication on the label, the new label should carry a date no later than the original durability indication.

22   WASTE MANAGEMENT

Principle

In the Introduction, the intention was made clear to limit the Guide to matters having a direct bearing on the scientific, technological and organisational aspects affecting quality, legality and safety of products. For this reason, detailed consideration has not been given to the impact of the manufacturing unit and its operations on the external environment. It is, however, acknowledged here that the management of any food manufacturing operation has general responsibility and, in most countries, legal obligations (with which it must be familiar) for these aspects including the management of waste.

General

22.1   **EU Regulation 852/2004** on the hygiene of foodstuffs and **Regulation 853/2004** laying down specific hygiene rules for food of animal origin came into force in all Member States on 1 January 2006. They are enforced by the **Food Hygiene (England) Regulations 2005** in England and by similar legislations in Scotland and Wales. These require:

(a) food waste, non-edible by-products and other refuse are to be removed from rooms where food is present as quickly as possible, so as to avoid their accumulation;

(b) food waste, non-edible by-products and other refuse are to be deposited in closable containers, unless food business operators can demonstrate to the competent authority that other types of containers or evacuation systems used are appropriate. These containers are to be of an appropriate construction, kept in sound condition, be easy to clean and, where necessary, to disinfect;

(c) adequate provision is to be made for the storage and disposal of food waste, non-edible by-products and other refuse. Refuse stores are to be designed and managed in such a way as to enable them to be kept clean and, where necessary, free of animals and pests; and

(d) all waste is to be eliminated in a hygienic and environmentally friendly way in accordance with the community legislation applicable to that effect, and is not to constitute a direct or indirect source of contamination.

22.2   The treatment/disposal of unwanted by-products (waste materials) must be controlled and comply with all the appropriate environmental legislation in the country concerned (see Chapter 30). In the UK, waste disposal should meet all relevant national and EC legislation. Where necessary, waste should be removed by licensed contractors. If the waste includes trademarked materials, the waste contractor should provide records of material destruction or disposal.

Food & Drink – Good Manufacturing Practice: A Guide to its Responsible Management, Sixth Edition.
The Institute of Food Science & Technology Trust Fund.
© 2013 John Wiley & Sons, Ltd. Published 2013 by John Wiley & Sons, Ltd.

22.3 The **EC Directive on Packaging Waste (94/62/EC)** aims to reduce the volume of packaging waste going to landfill sites by setting targets for the recovery and recycling of packaging waste. The UK legislation that implements the EC directive is the **Producer Responsibility Obligations (Packaging Waste) Regulations 1997**. The aim of the legislation is to ensure that those who produce or utilise the packaging, carry the real environmental costs of their actions and the means of disposal.

22.4 Waste management procedures should address the requirements for waste minimisation, reusing the material wherever possible, waste recycling using approved contractors and waste disposal. The frequency of emptying of waste containers should be defined, and verification activities should be undertaken to ensure that the timescales are complied with.

22.5 Where licensed contractors are required by legislation for the carriage and disposal of waste from the manufacturing site, all appropriate documentation should be maintained.

22.6 External waste areas should be maintained in an appropriate hygienic state. Internal and external waste containers should be clearly identified, that is, with labels or colour coding for different types of waste and designed and maintained so that they are fully enclosed are effective in use and designed in such a way that affords effective cleaning and, where required, disinfection (see Chapter 14). Consideration should also be given to effective infestation control (see Chapter 15).

Waste Minimisation Audit

22.7 Waste minimisation audits should identify ways to prevent waste being produced in the manufacturing unit. They should address the following areas as appropriate:

- raw material use and actual versus expected yield;
- level of non-conforming/rejected product;
- level of waste by-products produced per unit of production;
- potable water consumption and the volume of waste water produced per unit of production;
- energy consumption per unit of production;
- packaging usage and volume of packaging waste produced per unit of production;
- chemical usage; and
- sundry and consumable items.

23   WAREHOUSING, TRANSPORT AND DISTRIBUTION

Principle

Ensuring the safety, legality and quality of food in storage and in the distribution supply chain requires the development of procedures to ensure the preservation of food and to minimise the risk of contamination. Verification procedures also need to be in place to ensure that operating procedures are complied with and are effective.

General

23.1   Effective warehousing operations should be designed to ensure that all products are easily accessible for load assembly as required; to ensure that aisles and assembly areas are planned so that unimpeded movement is possible to and from all parts of the warehouse; to facilitate proper stock rotation, particularly important in relation to short-life and date-marked foods; and to obtain maximum utilisation of available space, consistent with the foregoing requirement. Design of premises should provide separate routes of entry and movement for vehicles and personnel. Designated walkways should be marked in internal and external areas. Vehicles should be loaded and unloaded in a manner that protects the material and/or product being transported. Where the material is susceptible to temperature abuse and/ or weather damage, loading and unloading should be undertaken in covered bays.

23.2   Storage and transport of finished products should be under conditions that will prevent contamination, including development of pathogenic or toxigenic microorganisms, will protect against undesirable deterioration of the product and the container, and will assure the delivery of safe, clean and wholesome foods to consumers. This deterioration includes, but is not limited to, contamination from insects, rodents and other vermin, toxic chemicals, pesticides and sources of flavour and odour taint.

23.3   The buildings, grounds, fixtures and equipment of food warehouses and vehicles should be designed, constructed, adapted and maintained to facilitate the operations carried out in them and to prevent damage. For practices of a general nature referring to these requirements, see Chapters 5, 6 and 12.

23.4   Pallets should be placed in prescribed places and stored under cover; gangways should be used as such and not as temporary repositories for stocks.

23.5   Product stacking should have regard for all elements of safety. Pallets should be checked periodically for structural integrity especially incoming and outgoing goods. Cornerboards should be positioned at the corner of each stack, both to make the corner stand out and to protect the product from accidental impact damage by high lift and

Food & Drink – Good Manufacturing Practice: A Guide to its Responsible Management, Sixth Edition.
The Institute of Food Science & Technology Trust Fund.
© 2013 John Wiley & Sons, Ltd. Published 2013 by John Wiley & Sons, Ltd.

powered pallet trucks. Palletised product should be checked for stability and that there is no product overhang. Shrink wrap may also be used to minimise product movement on the pallet, but care should be taken not to crush the food products through the tension of the shrink wrap. Pallet stacking configuration and pallet labelling should conform to either internal or customer specifications and should be routinely monitored by quality control.

23.6 When assessing the suitability of wooden pallets for use, the following should be considered:

- design and dimensions are fit for purpose;
- maximum weight load;
- signs of damage and the need for repairs;
- use of any preservation materials that could taint the product;
- dryness of the wood;
- missing blocks;
- protruding nails that could cause damage or affect packaging integrity; and
- potential pest contamination of the pallets.

23.7 Pallets of product should be so spaced as to allow proper ventilation.

23.8 A suitable curtain should be provided at all entrances and exits in order to maintain the internal conditions of the warehouse at an appropriate level for the products therein. A risk assessment should be undertaken, which is formally recorded, that identifies whether covered bays for vehicle loading are required. The use of roller doors, sliding doors or strip or air curtains should be assessed and appropriate measures adopted (see 12.11).

23.9 Warehouse and loading dock temperatures, particularly those for chilled or frozen food storage areas should be kept at an appropriate level to maintain the wholesomeness of the particular foods received and held in such areas. Routine monitoring of temperature should be undertaken to ensure that storage temperatures remain within defined limits. Temperature monitoring records should be maintained. Procedures should be documented and implemented, which define the actions to take in the event of a breakdown in the store.

23.10 The lighting should be as high as possible above the product; the smaller the angle of light source from ground level, the smaller is the shadow made by the stack. All glass and hard plastic items should be suitably protected as per the brittle material control procedure (see Chapter 12, Section 12.36–12.42).

23.11 As soon as product damage occurs or is discovered, damaged goods should be placed in a designated area. Consideration should be given as to whether contamination of other products has been possible, for example, if stored above other materials in racking when damage occurs. This is particularly important in the event that allergenic materials are being stored or transported. Care must be taken not to expose foods stored in the warehouse to contamination or infestation. Returns from customers must be assessed for contamination or infestation before being placed in a storage area. Returns from customers must be placed in a designated area until a formal review of disposition is undertaken by the quality control manager or designate (see Chapter 20, Sections 20.10–20.12).

23.12 Only products that have been properly inspected to ensure that the product and packaging are fully acceptable may be repacked into outer packaging. If it is necessary to repack goods of different production codes into the same outer packaging, the package should be marked with an age code that relates to the oldest packet in the case, that is, the shortest product duration.

23.13 Damaged goods that cannot be repacked must be dealt with prior to disposal so as to prevent their re-entry into the food distribution chain.

23.14 Docks, railway sidings, bays, driveways and so on should be kept free from accumulation of debris and spillage.

23.15 Fire exits should be checked on a routine basis to ensure that they are kept clear and allow access. Fire appliances should be suitable for use on the commodities concerned and a sufficient proportion of them should be capable of dealing with electrical and petroleum/fuel oil fires. Fire appliances should be checked by a suitably qualified individual or contractor at least annually. Certificates of inspection should be maintained.

23.16 Forklift and other trucks used within the warehouse should normally be battery driven or otherwise equipped to prevent fume or fuel contamination. Procedures should be in place to undertake regular inspections of forklift trucks to minimise the potential for product contamination and records should be retained of the inspections undertaken. Only competent personnel with suitable training should drive the forklift trucks.

23.17 Vehicles, particularly those used for the transport of chilled or frozen foods, should be capable of achieving, and be operated at, temperatures appropriate to maintain the wholesomeness of the foods being carried therein. Routine monitoring of temperature should be undertaken to ensure that vehicle temperatures remain within defined limits. Temperature data logging equipment should be used where deemed necessary to demonstrate that the vehicle temperature has remained within specified parameters during distribution. Temperature monitoring records should be maintained (see 33.22). Vehicle procedures should document and implement:

(a) maintenance and hygiene procedures including checks of vehicles to identify any potential source of odours that could taint product (see 23.19). The requirements for designated loads and the restrictions on mixing of loads should be defined;

(b) actions to be taken to ensure security of the load and in the event of a mechanical or temperature breakdown; and

(c) records that need to be maintained.

When developing hygiene procedures for vehicles, the cleaning of hoses and coupling points must be addressed and hoses should be capped when not in use. Care must be taken to ensure that during loading/unloading, the cap is not in contact with the ground or otherwise contaminated. With bulk deliveries, and where required by the manufacturer or customer, records should be retained of the last three loads that have been carried on the trailer/tanker and the cleaning that has been undertaken. The haulage company must comply with any haulage exclusion policies.

23.18 All vehicles, containers and so on should be free from rodents, birds and insects or contamination from them; free from odours, nails, splinters, oil and grease, and accumulations of dirt and debris; and should be in good repair, without holes, cracks or crevices that could provide entrances or harbourages for pests.

23.19 Prior to loading, the vehicle interior (including walls, floor and ceiling and light(s) if internal lights are present) should be inspected for general cleanliness, freedom from moisture, foreign materials and so on, which could cause product contamination or damage to the packages. Lights should be checked to ensure they are intact. The inspection should be formally recorded and linked to the registration number of the vehicle to ensure traceability of information as well as the delivery note (where the vehicle number should also be recorded).

23.20 The load of goods must be evenly distributed so as not to cause the gross weight, or any one of the axle weights of the vehicles, to be exceeded. Where a vehicle is loaded with multiple drops for distribution centres, a load plan should be completed that identifies the position on the vehicle of each pallet. One copy should travel with the vehicle, and one copy should be retained for the quality records.

23.21 The load should ride satisfactorily and safely when on the move in order to avoid damage to packages, people or the vehicle in the event of violent braking or cornering.

23.22 Vehicles bringing product to a warehouse should be inspected before offloading for evidence of damage, or of insect or rodent infestations, objectionable odours or other form of contamination. The checks should include that any lighting or other brittle item is intact.

23.23 If damaged product is accepted and offloaded from the vehicle, it must be kept separate from other product and handled in a manner that will not expose other foods on the vehicle, or subsequently the food warehouse or storage area, to contamination or infestation.

23.24   A procedure should be set up to deal with consequences of accidents and damage occurring when goods are in storage or distribution, for example, spillage procedures, salvage or condemnation following damage to goods in a road traffic accident.

23.25   Security precautions should include means of preventing and deterring any tampering with goods in storage and distribution such as the use of tamper-evident seals. Increasingly, digital photography is being used to demonstrate security on release by the supplier before entering the distribution chain. These photographs are then emailed to the manufacturer prior to arrival. The delivery can then be checked on arrival at the manufacturer for signs of tampering.

23.26   Where quality activities are being undertaken during food distribution such as fruit ripening or meat maturation, the quality control manager must develop documented procedures and associated forms to ensure that the process is suitably controlled and that an effective inspection and sampling plan has been developed for both in-transit activities and also for inspection on arrival at the manufacturing site.

23.27   Where warehousing is contracted out, the premises, vehicles and conditions should be subject to the manufacturer's food control checks and assessed under the supplier assessment and performance monitoring procedures.

23.28   Where contracted transport is used, documented procedures and/or terms and condition should be implemented and performance should be assessed under the supplier assessment and performance monitoring procedures. Further information can be found at the International Cold Chain Technology (ICCT) web site (http://www.icct.org.uk).

23.29   In the event of an outbreak, or a suspected outbreak, of a notifiable disease such as foot-and-mouth disease, avian influenza, Newcastle disease or swine fever, current legislation must be complied with throughout the distribution chain including compliance with movement restrictions placed in surveillance zones.

24   CONTRACT MANUFACTURE

Principle   *Where complete or part manufacture is carried out as an own-label, private-label, distributor's-own-brand, contract packing or similar operation, the obligation is on the contract acceptor (the actual manufacturer) to ensure that outsourced processing and production is carried out in accordance with good manufacturing practice (GMP) in the same way that would be expected where (s)he manufactures for distribution and sale on his/her own account, except where responsibility is specifically excluded by mutual agreement between the contract giver and the contract acceptor.*

Contract Acceptance   24.1   The contract acceptor should ensure that the terms of the contract are clearly stated in writing and that raw materials, in-process materials and end products are covered by adequate, comprehensive specifications (as outlined in other chapters). Any special GMP requirements should be clearly emphasised, and quality control, record transfer, coding, rejection, dispute and complaint procedures should be identified and agreed. Items of possible confidentiality should be identified and any appropriate safeguards be mutually agreed.

24.2   If the contract giver has a designated approved suppliers list, this must be discussed and agreed with the contract acceptor. The responsibility for approval of packaging and artwork design with the packaging supplier must also be agreed.

24.3   It is normal practice for contract givers to impose contractual conditions that ensure compliance with food safety, legal and quality standards and principles of GMP. This is desirably achieved, at least in the first instance, by a visit to the manufacturing unit, by the contract giver's quality control manager and/or brand owner/retailer product technologist as appropriate to the status of the contract giver. The visit should include the following objectives to:

(a) ensure that within the manufacturing environment, the food can be produced safely and legally, and to the required quality standards;

(b) agree a detailed product specification covering all aspects of the product, processing and packing requirements as well as controls for delivery, embracing parameters to be used for acceptance or rejection, and any legal requirements relating thereto;

(c) agree levels of sampling of raw materials, in-process materials and finished products by the processor and sampling plans to be used in case of dispute;

(d) evaluate the adequacy of the control resources, systems, methods including traceability and records of the manufacturer; and

Food & Drink – Good Manufacturing Practice: A Guide to its Responsible Management, Sixth Edition.
The Institute of Food Science & Technology Trust Fund.
© 2013 John Wiley & Sons, Ltd. Published 2013 by John Wiley & Sons, Ltd.

(e) agree, wherever possible, objective methods of examination, while subjective measurements should conform to recognised and accepted standards if possible.

Agreement in all five areas is essential for any manufacturer/customer trading relationship and should benefit both parties.

24.4   In the event that the contract giver is outsourcing part of the manufacturing of a branded product, then the brand owner should be notified and approval sought where required. The contract giver should also establish validation, monitoring and verification activities both prior to and for the duration of the contract with the contract acceptor. The extent of these activities will be based on a formal, documented risk assessment and may include but are not limited to organoleptic, chemical and microbiological assessment. The contract acceptor will be approved and monitored in line with the contract giver's supplier approval and performance monitoring procedure (see Chapter 16).

Third-Party
Certification

24.5   Historically, contract givers undertook the verification of manufacturing units against their individual company-specific standards, protocols and codes of practice. Indeed this is still the practice for some contract givers who have developed their own bespoke manufacturing standards. However, increasingly verification is being undertaken by third-party certification bodies especially as the supply chain becomes more global. The development of technical standards such as British Retail Consortium (BRC) Global Standard for Food Safety, BS EN ISO 22000, and latterly the work of the Global Food Safety Initiative (GFSI) has seen a rise in third-party certification. This certification is increasingly becoming a prerequisite to supply. It should be noted that acceptance of third-party certification, or the use of third-party auditors, as the sole evidence of suitability to pack or manufacture food products, renders the contract giver potentially unable to provide an adequate due diligence defence. There is a requirement for the contract giver to verify that third-party certification and/or third-party inspections and audits are undertaken to the required standard and that the decisions made by such auditors are appropriate and reflect actual practice. In this context, contract givers must determine the level of verification risk (VR) of accepting third-party certification and auditing/inspection. As VR increases, so ultimately does the risk of customer complaint, product withdrawal and recall and incidence of food-borne illness and food poisoning.

24.6   The evaluation for these aforementioned private standards must be undertaken by certification bodies who are accredited against the European Standard EN 45011 (ISO/IEC Guide 65) (General requirements for bodies operating product certification systems).

25   CALIBRATION

In order to ensure the validity of measuring inspection and testing equipment (MITE) that is being used to verify that the product is within specification, safe and meets current legislation and/or processing equipment is maintaining or monitoring process conditions, it should be routinely calibrated to a recognised national standard. During the validation process, the location and type of MITE and the required degree of accuracy should be determined, documented and then monitored and verified to ensure that process and product parameters are continuously maintained within designated boundaries. This is especially important at process steps that are identified as critical control points (CCPs) within the hazard analysis critical control point (HACCP) plan. Equipment calibration is therefore a key prerequisite within a food safety management system (FSMS) and quality management system (QMS) that incorporates the requirements of good manufacturing practice (GMP).

General

25.1   Calibration procedures should ensure that all equipment required for measuring, inspection, testing and process control is suitable to demonstrate conformance to specified requirements, regularly calibrated if necessary and its use(s) specified. Calibration procedures should also ensure that there will be no reduction in product quality in the event of failure of any item of inspection or test equipment. The quality control manager should be responsible for the effective design and implementation of equipment calibration procedures. These procedures should identify the actions to be taken in the event that calibration activities indicate that unsafe, illegal or out-of-specification product may have been manufactured and distributed. All calibration activities should be traceable to recognised national standards. The need for equipment sanitisation following calibration activities should be considered and appropriate procedures put in place to minimise the risk of product contamination.

25.2   Equipment may include, but is not limited to:

- weighing scales, weigh cells or checkweighers;
- standard test weights;
- counting devices, for example, magic eyes;
- flow meters;
- measuring equipment, for example, callipers and sizing rings;
- temperature probes;
- metal detection equipment;
- penetrometers;
- refractometers; and
- pH meters.

Food & Drink – Good Manufacturing Practice: A Guide to its Responsible Management, Sixth Edition.
The Institute of Food Science & Technology Trust Fund.
© 2013 John Wiley & Sons, Ltd. Published 2013 by John Wiley & Sons, Ltd.

25.3   Calibration may be either an internal activity undertaken by trained personnel or an external third-party calibration service. A MITE Register or equivalent document should be maintained for all items of inspection, and test equipment other than those that are 'for indication only'. The MITE Register should contain the following information as applicable:

- equipment type;
- serial number;
- internal company number;
- equipment location; and
- date last calibrated and date next calibration is due.

For each piece of equipment that requires calibration, a calibration record should be maintained detailing as applicable:

- details and description of the equipment;
- serial number or identification number;
- acceptable degree of accuracy required;
- the range in which the piece of equipment should be calibrated. The calibration range must reflect the range of temperatures, weights and so on within which the measuring and inspection and testing equipment will be required to perform during routine monitoring and verification activity;
- the location of the equipment;
- the procedure for calibration (internal or external);
- frequency of calibration;
- the next due date for calibration; and
- evidence of calibration—a certificate that is traceable to national standards.

The frequency of calibration and the degree of accuracy required when calibration is undertaken should be determined by risk assessment, which is formally documented and routinely reviewed to ensure that product safety, quality and legality can still be maintained. Work instructions and/or task procedures should be developed that outline the protocol for internal calibration activities. Only trained competent staff may undertake internal calibration activities, and training records should be available that demonstrate the level of training of staff. Calibration records should be routinely verified by the quality control manager or designate to ensure that all:

- calibration is up to date;
- equipment is working to the required level of accuracy; and
- equipment currently in use has a valid calibration certificate and is traceable to a recognised national standard.

A calibration diary is a useful tool to ensuring equipment is calibrated on time, especially where there is both internal and external calibration undertaken.

25.4   Once a piece of equipment has been calibrated either internally or externally, it should only be adjusted by authorised personnel and according to prescribed procedures. Any adjustments should be formally recorded. The equipment should bear a formal identification mark, stamp or serial number that is traceable to a calibration certificate and the aforementioned MITE Register. All internal weights that are used for daily calibration of weighing scales and checkweighers should be individually marked and traceable to calibration certificates and the aforementioned MITE Register.

25.5 Internal calibration and any required actions should be formally recorded. All MITE should be suitably protected from any damage, deterioration or misuse during handling, maintenance or storage.

25.6 All items that require calibration for which current calibration certificates or their equivalent are not available should be clearly marked as such, for example, 'Not calibrated, not in use'. Any equipment marked as such should not to be used for any inspection or testing purposes.

25.7 Equipment that is calibrated independently for which there is no back-up should be calibrated when the company is not in operation (i.e. weekends, holiday).

25.8 Equipment failing calibration should be repaired or replaced. Repaired items should then be recalibrated before reuse. In the event of any item failing calibration, the quality control manager should assess the validity of previous inspections.

25.9 The quality control manager should ensure that there is back-up equipment or an alternative procedure ready for operation in the event of any failure of inspection and test equipment. Back-up equipment should also be incorporated into the MITE Register and maintained in working condition.

25.10 The quality control manager should develop procedures to ensure that in the event that inspection and testing equipment is found to be functioning incorrectly, or damaged, all product produced since the last satisfactory check can be identified, isolated and retested. This will be a factor in determining the frequency of calibration of items of equipment. The need for product withdrawal/recall should be assessed where appropriate, and action taken to control any affected product (see Chapter 19).

25.11 All scales and checkweighers should also be calibrated and serviced routinely by an external contractor. The contractor should be monitored as per the supplier approval and performance monitoring procedure (see Chapter 16). Where automatic checkweighers are used, it should be verified at the start of production that they will reject low weight packs and then monitored at routine intervals during the production shift. If the automatic checkweigher fails to reject the low weight samples, the quality control manager should be informed and appropriate action taken. It is important that the low weight samples are identified in such a way that they cannot be confused with actual production packs.

25.12   If there is a failure during routine monitoring of the checkweigher, a back-up manual system of checkweighing must be developed so that it can be introduced until the equipment can be serviced and checked by an appropriately qualified person. All production packed since the last satisfactory check should be identified and rechecked manually to ensure its conformity with specifications. Any out-of-specification product must be suitably controlled.

26   PRODUCT CONTROL, TESTING AND INSPECTION

Principle

Ensuring the safety, legality and quality of products requires the development of testing and inspection procedures to ensure that the food products are safe, wholesome and comply with relevant legislation and meet consistently designated specifications. Testing and inspection procedures should enable the relevant parameters to be monitored so that corrective action can be taken if results fall outside specified limits.

General

26.1   Quality control staff need to be aware of the specified control limits for the raw material, ingredient or product being tested. Good laboratory practice is addressed in Chapter 29. Any external laboratory services used must be monitored as per the supplier approval and performance monitoring procedure (see Chapter 16).

26.2   Quality control personnel should have the authority to hold product considered to be outside specified control limits (see 11.9). They must then refer the matter to the appropriate manager for his/her formal decision on disposal.

26.3   Quality control checks should be scheduled throughout the process including intake, on-line and despatch checks. Quality control records must be completed for all batches/consignments and should be held on file for both internal reference and customer inspection. Historical quality control records should be held according to the controlled records list (see 9.9). Any changes to quality control records must be authorised and the reasons for the change recorded and who has approved the change.

26.4   Quality control procedures should be developed by the quality control manager considering the following:

- product characteristics to be assessed and the acceptance criteria;
- the 'volume' of product that constitutes a batch;
- sampling accuracy and whether the sample is indicative of the whole batch;
- equipment required and the appropriate measurements to be taken and the degree of accuracy required for the equipment;
- skills, training and qualifications required by quality control personnel;
- points in the process where the measurements will be undertaken and the frequency of checks;
- actions to be taken in the event of non-conformance;
- the documentation to be completed that records the results of inspection; and
- the requirements for verification of the results of product testing.

Food & Drink – Good Manufacturing Practice: A Guide to its Responsible Management, Sixth Edition.
The Institute of Food Science & Technology Trust Fund.
© 2013 John Wiley & Sons, Ltd. Published 2013 by John Wiley & Sons, Ltd.

26.5   Product characteristics/criteria to be assessed at each stage should be defined in a quality plan. Product characteristics/criteria may also be described in individual customer specifications and could include:

- sample size. This may be determined by legislation, customer requirements or internal procedures. It is important to ensure that the sampling is large enough to be representative of the batch being tested;
- product temperature;
- traceability as required by the specification;
- origin, livestock breed and variety. If there are varietal differences such as with produce, especially where there is a legal requirement to label the product with varietal declaration, this must be adequately controlled. The material must be checked to ensure that it is from an approved source, is of the correct provenance if this has been determined as a requirement and is as specified in accompanying documentation. If the material is specified as being from a certified source, for example, organic, kosher and farm assured, then identity preservation must be maintained through the manufacturing process;
- quality and appearance. The product must be assessed as per internal/customer specification including as appropriate intrinsic standards such as smell, colour, size and shape, maturity, eating quality, freshness, firmness and cleanliness or level of soiling;
- specific product characteristics that are applicable to individual ingredients and products, for example, pH, water activity (a_w), sugar content (% Brix), salt content and preservative levels; and
- packaging in terms of compliance with specification, signs of damage, correct labelling and product information (see 26.11).

26.6   Any product failing to meet specifications must be placed on hold and suitably marked and/or held in a quarantine area awaiting a decision on its disposition.

26.7   If the product fails to meet the specification at any stage in the process, then it should be held, and the quality control manager informed. (S)/he should decide, with other members of the management team as necessary, the final product disposal. Disposal will be either to reject, to use under concession, to supply to an alternative customer whose specifications it meets or to reprocess or rework material. The appropriate action should be taken only on the written authority of the quality control manager. Causes of non-conformance should be recorded and analysed and corrective action initiated to prevent recurrence. Corrective action needs to be documented and followed up to ensure that it has been completed and is effective (see Chapter 20).

Quantity Control   26.8   Quantity control systems could be based on average quantity, or minimum quantity. The procedure for quantity control should meet the minimum requirements of legislation and also the customer requirements as defined in the product specification. Quality control and production personnel must be aware of the minimum gross pack weight/volume required for each individual pack/production line (including stated product weight/volume, packaging tare weight and any weight loss allowance for product that could dehydrate over its shelf-life duration). Packed product may also be sold to a specific count. Where bulk product is not subject to legislative requirements, it must conform to the relevant specification.

26.9 All weighing scales and weighing equipment should be tested daily against standard test weights, which have themselves been calibrated and are traceable to national standards. The results should be recorded on the appropriate record (see Section 25.11). Before commencing work, the production and/or quality control personnel, as applicable, should ensure that the equipment they are using is working correctly and suitable for use. At the end of the production shift, they should ensure that the equipment is left in good working order. In the event that equipment is broken, the quality control manager, or designate, should be informed so that (s)/he can arrange for the equipment to be repaired and determine the need for product to be held, withdrawn or recalled.

26.10 The use of statistical process control (SPC) especially with automatic filling machines or checkweighers can assist in the minimisation of 'giveaway' and ensure effective control of product weights and thus productivity. The roles and responsibilities for quality control and production personnel in implementing the quantity control systems need to be formally defined and routinely assessed to ensure compliance with legislative requirements and product specifications.

Label Control   26.11 Product labels should be kept in a designated area until required by production. Labels should be issued from this area to authorised personnel, and any unused/returned labels should be signed back in and accounted for at the end of production (see Chapter 27). During product changeovers, all packaging and labelled materials especially flash or promotional labels should be removed from the line and the line should be approved by the line supervisor or designate that it is ready to start manufacture of the next product. In the event that labels from previous production runs are found on the production line, the quality control manager should be informed immediately and appropriate action taken including the potential withdrawal of product from the distribution system. All actions taken should be documented.

26.12 Quality control personnel should approve labels prior to use. The label should be checked by at least two and preferably three competent individuals before it is approved for production. This should be recorded on the appropriate quality record. The quality checks should include as applicable to the label and relevant product specification:

- label format (portrait or landscape) and position on the pack especially in relation to additional flash or promotional labels;
- product label design and descriptions including brand owners name and address, quantitative ingredient declaration, nutrition labelling and claims;

- print registration and quality;
- colour coding (where applicable e.g. on box end or tray labels);
- product, variety and country of origin;
- volume, weight, size or count;
- bar code (readability to be confirmed by use of an approved bar code scanner and verifier);
- customer department/commodity number (where applicable);
- processing plant/supplier identification number/code;
- durability date either use by, display until or best before, or a combination of the latter two, depending on specific customer requirements;
- price (where applicable); and
- print clarity through the production run and correct size of print.

The first and the last label of both product, pack and tray end label should be checked from each batch (reel) of label, signed off and affixed to the quality control record. If film is printed on-line, then samples of printed film should be retained. Labels should be checked at designated intervals during the production run for continued conformance. The frequency of checks could relate to a period of production time or production volume, for example, every 30 minutes, every hour or every pallet.

Shelf-Life Assessment 26.13 Samples should be held for the designated time period and tests carried out as detailed in individual internal/customer specifications. If any rapid deterioration of the product occurs, this should be referred to the quality control manager or designate so that appropriate action can be taken. Shelf-life samples may be held at different temperatures including frozen, chilled, ambient or accelerated storage. Shelf-life records should be retained and should be analysed for potential trends.

Despatch Quality Control Procedures 26.14 Quality checks and the temperature of all products must be taken prior to despatch and noted on the appropriate record. If the finished product is subject to positive release checks, then these must all be undertaken, recorded and signed off if fully compliant so that the batch/load can be formally released for distribution. Positive release procedures should be documented and implemented, and only authorised personnel may formally release a product as specified in the procedure(s). The traceability information on these records should concur with those on delivery and despatch notes.

Food Safety Monitoring and Verification 26.15 Food safety monitoring and verification activities will be undertaken in line with the prescribed requirements of the food safety management system (FSMS) and/or hazard analysis critical control point (HACCP) plan. The testing undertaken at both control points and critical control points (CCPs) should demonstrate that the process or product complies with the required critical limits and or target levels and tolerances where these have been defined (see 3.13). Verification activities will also include product inspection and testing activities such as shelf-life testing, testing of products from the retail shelf and other tests as identified by the FSMS.

27   LABELLING

Principle   *The labelling and presentation of the product, which includes any informa-*
tion printed on or implied by the shape or design of the packaging or by the
application of labels to the packs, must correctly correspond to the product
concerned, and must comply with the legislative requirements of the country
in which the product is to be sold.

General   27.1   In many manufacturing units, there will be some combination of
different products, different sizes of a given product and different
versions of a given product (whether variations in composition or in
branding or in language), and it becomes essential to ensure that the
correct supply of packaging and/or label is selected. The correctness
of selection should be independently checked by the production
management team and quality control before a production run, and
any residual material from a previous production run removed from
the production area (see 26.12). This should be controlled by formal
documented procedures.

27.2   When a new pack or label design is introduced for a product, residual
stocks of the obsolete packaging or label should be destroyed at the
designated changeover time. Similarly where packaging or labels are
reference coded or date marked before use, surplus material left from
earlier production and bearing a no longer valid reference or date
should be withdrawn from the production area and destroyed. Redundant packaging can incur significant costs and needs to be effectively
managed.

27.3   A particular source of selection error may be the circumstance of
material being packed and/or processed in unmarked containers,
which are then stored unlabelled for a period of whatever length
before being labelled or placed in further containers carrying the
required information. All such stored material should be identified
with full information to enable it to be correctly labelled in due
course, with adequate precautions to ensure that its full identity does
not become lost (see Chapter 10).

27.4   The task of designing packaging or labels that, while fulfilling
the presentation wishes of the manufacturer, also comply with the
requirements of the law of the country for which the product is
intended, is a complex and difficult one. Legislative requirements
governing the labelling of a product are rarely to be found in a single
law. In addition to general labelling regulations, the compositional
regulations for some products contain some special labelling requirements, which sometimes replace and sometimes add to the provisions
in the general labelling regulations. Some aspects of labelling

Food & Drink – Good Manufacturing Practice: A Guide to its Responsible Management, Sixth Edition.
The Institute of Food Science & Technology Trust Fund.
© 2013 John Wiley & Sons, Ltd. Published 2013 by John Wiley & Sons, Ltd.

requirements are also governed by weights and measures legislation and various consumer protection laws. In addition to regulations, account needs to be taken of case law on the way in which regulations have been interpreted by courts, guidance from official sources and relevant industry codes of practice.

27.5   For a particular product, account has to be taken of all the different requirements affecting its labelling and packaging, not only in terms of the information included but also in terms of its spatial layout, presentation and design. In many cases, interpretations have to be made of how the relevant provisions apply in the particular combination of circumstances presented by the product. Additionally, changes are continually taking place in labelling legislation in most developed countries (and some developing ones), and there are usually transitional periods when the old and the new regulations operate side by side, requiring detailed knowledge of the changes and the precise timetable. It will be seen that ensuring compliance is not a task that can be left to graphic designers or advertising agencies or marketing managers or product managers, but requires a considerable degree of right-up-to-the-minute technico-legal expertise. Such expertise is an essential part of food control and within the role of the quality control manager with such assistance as (s)/he may require.

27.6   When a new label for a product is being designed, or an existing one is being revised, whether expert help is sought from inside or outside the company or both, that label must comply with the legal requirements of the country in which the product is to be sold. Final adoption of the label should not be made without the approval of the quality control manager.

27.7   In the UK and the other European Union (EU) Member States, most food labels are legally required to carry a durability indication giving 'the date up to which the food can reasonably be expected to retain its specific properties if properly stored' plus a statement of 'any storage conditions which need to be observed if the food is to retain its specific properties until that date'. This necessitates the manufacturer establishing for each product a shelf life, that is, the duration from the date of manufacture that determines the date to be given on the pack.

27.8   Determining the durability indication on labels for a particular product involves considerations of microbiological safety; of retention of acceptable organoleptic properties as assessed by sensory analysis; of chemical, physical and microbiological characteristics; and, in the case of nutrition information including vitamin declaration(s), compliance with the amounts declared. It is not enough to guess at the shelf life, or to use what is thought to be (or worse, longer than is thought to be) the shelf life used by another manufacturer for an apparently similar product.

27.9   Proper scientifically based testing should be carried out to establish the shelf life to be used for a product in a specific packed format.

For details of the circumstances relating to the use of 'best before' and 'best before end', and the circumstances that trigger the requirement for a 'use by' declaration, see **Directive 2000/13/EC** as amended, and in the UK, the **Food Labelling Regulations 1996, SI 1996/1499. These regulations are amended by a series of subsequent legislation that is identified in the Food Law Guide, which can be accessed at** http://www.food.gov.uk/foodindustry/regulation/foodlaw/foodlawguide/.

27.10 The **Food Labelling Regulations 1996, SI 1996/1499** also addressed Quantitative Ingredient Declarations (QUID) labelling.[1] This labelling requirement applies to the labelling and marking of pre-packed food that is sold direct to the consumer/customer in its current packaging. The QUID requirement designates the circumstances in which the percentage(s) of one or more ingredients in a product must be declared in the ingredients list. **Regulation EU No 1169/2011** on the provision of information to consumers was published in October 2011.[2]

While many provisions are carried forward from existing legislations, there are some important new requirements. The legislation shall apply on 13 December 2014, with the exception of one point that shall apply on 13 December 2016, and another that shall apply on 1 January 2014. Labels complying with the legislation before 13 December 2014 will be legal.

27.11 See Chapter 38 regarding labelling of irradiated foods, Chapter 39 regarding labelling of novel foods and genetically modified foods, and Chapter 4 regarding labelling requirements in respect of the presence of the legally specified food allergens.

---

[1]http://www.food.gov.uk/multimedia/pdfs/clearfoodlabelling.pdf.
[2]http://eur-lex.europa.eu/LexUriServ/LexUriServ.do?uri=OJ:L:2011:304:0018:0063:EN:PDF.

28 ELECTRONIC DATA PROCESSING AND CONTROL SYSTEMS

Principle

The use of electronic data processing (EDP), including the use of computers for storing and handling any kind of documentation referred to in Chapter 9, or the use of computers or electronic devices in control of processes or operations, does not in any way alter the need to observe the relevant principles set out in this Guide. However, it imposes the additional need for safeguards to ensure that there is no consequent adverse effect on the achievement of requisite product safety, legality or quality, and no risk of required records or data being irretrievably lost, damaged, corrupted or altered in other than a properly authorised way.

Implementation and Operation

28.1   Before a computer system is put into operation, the required purpose should be clearly defined, and it should be tested for capability to achieve that purpose. If a manual system is being replaced, or an existing system is being replaced by a new one, the old system should be continued alongside for a period, both as part of the validation of the new system and as a safeguard in the event of problems or teething troubles in the new system.

28.2   Access to data and documents should be readily obtainable by authorised persons where necessary.

28.3   An effective back-up system is essential (see 28.10).

Responsibility for Systems

28.4   The use of computers and electronic data control systems does not change the responsibilities, as set out elsewhere in this Guide, of key personnel but additionally requires close collaboration between them and those responsible for computer systems. Where this responsibility lies should be clearly stated. Persons using the EDP system should be appropriately trained in its use, and should be able to obtain rapid expert advice to deal with any problems that might arise in its use. The requirements outlined in Chapter 9 with regard to document and data control apply as equally to electronic data, as they do to data in paper format. This is especially so for the management of obsolete files, documents and data held electronically either in a memory that has personal access only or on a server. Consideration should be given to the use of read-only files by those who can access but do not have authority to alter or amend electronic documents such as policies, procedures and work instructions.

Security

28.5   **Where the system contains personal information (e.g. related to consumer complaints), governed by the UK Data Protection Act 1998, the requirements of the act must be observed**.

Food & Drink – Good Manufacturing Practice: A Guide to its Responsible Management, Sixth Edition.
The Institute of Food Science & Technology Trust Fund.
© 2013 John Wiley & Sons, Ltd. Published 2013 by John Wiley & Sons, Ltd.

28.6   Alterations to the system or a computer programme should be made only with proper authorisation, and in accordance with a defined procedure, which should include the checking, approving and implementing of the change. When any such change is made, previously stored data should be checked for accessibility and accuracy. A log of all such alterations should be maintained to record the date, change details, person making alteration, authorisation details and any other pertinent information.

28.7   By means of appropriate keys, pass cards, personal codes or passwords and restriction of access to computer terminals and recording media, the system should contain safeguards against unauthorised access to, or alteration of, any data. Personnel, visitor and contractor procedures should clearly outline the protocols that are in place for the ownership and use within the organisation of portable back-up hard drives, memory sticks and pens, laptops, tablets or other mobile electronic devices that are brought onto the premises. Where the computer is part of a network (whether internal or external), security measures against unauthorised access should be taken. Where the same equipment is used for other functions (e.g. accounting, sales records, personnel records), access to the types of data referred to in this chapter should not enable access to data pertaining to such other company functions. Data should only be obtained, entered or amended by persons authorised to do so, and there should be a defined procedure for the issue, cancellation or alteration of authorisation. Where critical data are altered by an authorised person, the record should show that the alteration has been made, by whom and the reason.

28.8   When critical performance data are being entered on a computer file record, there should be an independent check on the accuracy of the entry.

28.9   The system should be designed, precautions provided and controls (of the type mentioned in Section 28.10) exercised so as to safeguard against accidental or wilful damage of stored data by persons or physical or electronic means. A system of cross-checking for any loss of data is advisable.

Back-Up   28.10   There should be available adequate back-up arrangements for any system that needs to be operated in the event of a breakdown or emergency, and such back-up arrangements should be capable of being called into use at short notice. As a safeguard against loss or corruption of stored data and documents, provision should be made for back-up copies of data and systems software to be stored remotely from the computer's location; such copies should be kept up to date, and tested periodically. Access to the types of documents listed in Chapter 9 should be available to the quality control manager and the production manager.

28.11   The procedures to be followed in the event of a system failure or breakdown should be defined and routinely tested. Any failures, and remedial action taken, should be recorded.

28.12 If hardware service or software maintenance is provided by an outside agency, there should be a formal agreement including a clear statement of the responsibilities of that agency including a clause on the maintenance of confidentiality.

28.13 Consideration should be given to the ease of access to files with regard to system hacking, viral attack or other activity that could compromise the security of controlled data.

29   GOOD CONTROL LABORATORY PRACTICE AND USE OF OUTSIDE LABORATORY SERVICES

Principle

A control laboratory should have appropriate premises, facilities and staff, and be so organised, as to enable it to provide an effective service at all relevant times necessary to fulfil good manufacturing practice (GMP) requirements.

EU GLP

29.1   The European Union (EU) has published the acceptance of the Organisation for Economic Co-operation and Development (OECD) recommendation on compliance with the principles of good laboratory practice (GLP) (89/569/EEC). In addition, there is a Commission Directive (90/18/EEC) concerning the inspection and verification of GLP.

Resources

29.2   The resources required will depend on the nature of the materials and/or products to be tested. It is essential that facilities are appropriate to the needs of the tests, whether chemical, physical, biological or microbiological. Staff should be properly trained, well motivated and well managed. Standards should be set at the highest level and maintained by careful attention to approved and agreed methods and method checks using, where appropriate, reliable outside expertise.

Premises

29.3   Control laboratories should be designed, equipped, maintained and of sufficient space to suit the operations to be performed in them. This will include provision for writing and recording and the storage of documents and samples, and refrigerated storage for samples, as required.

29.4   Chemical, biological and microbiological laboratories should be separated from each other and from manufacturing areas. Separate rooms may be necessary to protect sensitive instruments from vibration, electrical interference, humidity and so on. Care should be taken to avoid contamination in either direction between laboratories (particularly microbiological laboratories, where access and exit controls should be strictly followed) and manufacturing areas, and reagents or materials that could cause taint should ideally be kept in a separate building. Provision should be made for the safe storage of waste materials awaiting disposal **(Control of Pollution Act 1974 and Environment Protection Act 1990)**.

29.5   **The UK legislation concerned with fire and explosion in laboratories is mainly contained in the Factories Act 1961 when laboratories are concerned with process control, and the Petroleum (Consolidation) Act 1928 and regulations and by-laws made under these acts. The Highly Flammable Liquids and Liquefied**

Food & Drink – Good Manufacturing Practice: A Guide to its Responsible Management, Sixth Edition.
The Institute of Food Science & Technology Trust Fund.
© 2013 John Wiley & Sons, Ltd. Published 2013 by John Wiley & Sons, Ltd.

Petroleum Gases Regulations 1972 (SI 1972 No 917) made under the Factories Act regulate the use and storage of such materials.

29.6 All services should be identified using colour coding according to standard documented procedures.

COSHH 29.7 **In the UK, the Control of Substances Hazardous to Health (COSHH) Regulations 1994, SI 1994 No 3246 as amended by SI 1996 No 3138 Amended 2004** affect the choice of safe laboratory working methods. All methods written up should include an assessment of the hazard of each of the chemicals used in the analysis and appropriate instructions to contain any hazard. If necessary, monitoring of the exposure to hazardous chemicals should be carried out. Advice on COSHH may be obtained (in the UK) from the Health and Safety Executive, among others.

Equipment 29.8 Control laboratory equipment and instrumentation should be appropriate to the testing procedures undertaken.

29.9 Equipment and instruments should be serviced and calibrated at suitable specified intervals by an assigned competent person, persons or organisation. Approved laboratories working to traceable national standards should in turn calibrate measuring inspection and test pieces used in the calibration process. Records of the calibration procedure and results should be maintained for each instrument or item of equipment or test piece. These records should specify the date when the next calibration or service is due (see Chapter 25).

29.10 Written operating instructions should be readily available for each instrument.

29.11 Where practicable, suitable arrangements should be made to indicate failure of equipment or services to equipment. Defective equipment should be withdrawn from use until the fault has been rectified. Following identification of defective or broken equipment, a review should be undertaken by the quality control manager or designate to determine the potential impact of equipment failure on product safety, legality or quality. Actions taken as a result of the review should be documented and verified to determine their effectiveness.

29.12 As necessary, analytical methods should include a step to verify that the equipment is functioning satisfactorily.

Cleanliness 29.13 Control laboratories and equipment should be kept clean, in accordance with written cleaning schedules. Adherence to cleaning schedules and compliance with cleaning standards should be verified at routine intervals. Verification activities should be recorded.

29.14 At all times, personnel should wear clean protective clothing appropriate to the duties being performed, especially eye protection.

29.15   The disposal of waste material should be carefully and responsibly undertaken by appropriate bodies.

Reagents, Controls   29.16   Where necessary, reagents should be dated upon receipt or prep-
and Standards   aration.

29.17   Reagents made up in the laboratory should be prepared by persons competent to do so, following laid down procedures. As applicable, labelling should indicate the concentration, standardisation factor, shelf life and storage conditions. The label should be initialled or signed, and dated, by the person preparing the reagent. A date for restandardisation should be recorded if required.

29.18   In certain cases, it may be necessary to carry out tests to confirm that the reagent is suitable for the purpose for which it is to be used. A record of these tests should be maintained.

29.19   Both positive and negative controls should be applied to verify the suitability of microbiological culture media. The size of the inoculums used in positive controls should be appropriate to the sensitivity required.

29.20   Reference standards, and any secondary standards prepared from them, should be dated, and be stored, handled and used so as not to prejudice their quality.

Sampling   29.21   Samples should be taken in such a manner that they are representative of the batches of material from which they are taken, in accordance with written sampling procedures approved by the quality control manager. The sampling activities used are constrained by:

- the resources and time available;
- the planned frequency of verification activities;
- the volume of data to be assessed;
- any planned or unplanned sampling bias; and
- the potential for deviation from the scope of the sampling protocol.

Sampling procedures should identify whether they are risk-based or non-risk-based methods of sampling, that is, whether all batches have an equal probability of being sampled or whether, though a risk assessment process using screening criteria, the sampling is biased towards certain ingredients, products and potential food safety hazards.

For example, materials associated with possible contamination such as mycotoxins (e.g. nuts) might be sampled at a higher frequency than those products that are not identified as having the potential to be so contaminated. Risk-based sampling accepts the premise that resources are limited and that sampling must be economically viable and deliver benefits to the manufacturing business. Other screening criteria might be third-party certification, provision of certificates of analysis, history of conformance with specifications and the volume of product utilised by the manufacturing business and thus the impact of failure.

The European Commission (1976[1] and 2006[2]) distinguished between different types of food samples namely:

- *Sampled Portion:* a quantity of product constituting 'a unit' having characteristics presumed to be uniform;
- *Incremental Sample:* a quantity taken from one point in the sampled portion, lot or sublot; and
- *Aggregate Sample:* an aggregate of incremental samples taken from the same sampled portion or the combined total of all the incremental samples taken from the lot or sublot.

Therefore it is important to classify the type of sample being derived as this will impact on the results in terms of the amount of variability in the batches to be sampled and whether this is masked by an aggregate sample and whether contamination in one batch is 'diluted' by the use of aggregate sampling to a point where it is not detected or finally where traceability of a food safety hazard identified during sampling to a particular batch is no longer possible. The physical state of the material, that is, liquid, solid, particulates, will also influence the sampling protocols that can be effectively adopted.

Sampling procedures should include the:

(a) method and rate of sampling. This should reflect the degree of homogeneity of the material being sampled;
(b) equipment to be used;
(c) amount of sample to be taken;
(d) instructions for any required subdivision of the sample;
(e) type and condition of the sample container to be used;
(f) storage requirements for the sample prior to testing;
(g) any special precautions to be observed, especially in regard to sterile sampling or sampling of noxious materials and the prevention of false positives; and
(h) cleaning and storage of sampling equipment.

Any sampling by production personnel should only be done by competent personnel in accordance with these approved procedures.

29.22 Each sample container should bear a secure and indelible label indicating its contents, with the batch or lot number reference and the

---

[1]European Commission (1976). First Commission Directive of 1 March 1976 establishing Community methods of sampling for the official control of feedingstuffs, 76/371/EEC. In: Official Journal, L 102, 15/04/1976, pp. 1–7.

[2]Commission Regulation of 23 February 2006 laying down the methods of sampling and analysis for the official control of the levels of mycotoxins in foodstuffs, 2006/401/EC. In: Official Journal, L 70, 09/03/2006, pp. 12–34.

date of sampling. It should also be possible to identify the bulk containers from which samples have been drawn.

29.23   Sampling equipment should be cleaned after each use and stored separately from other laboratory equipment.

29.24   Care should be taken to avoid contamination, or causing deterioration, whenever a material or product is sampled. Special care is necessary when resealing sampled containers to prevent damage to, or contamination of or by, the contents.

Methods   29.25   Methods should be chosen with care to fulfil the needs of the analyses. For quality control purposes, the chosen method should be that most efficacious for the accuracy and speed of results needed, and the skill of the staff concerned. When possible, methods acceptable to any enforcing authority, or which are internationally acceptable, should be used. In all cases, method checks need to be incorporated into any analytical scheme to ensure reproducibility, repeatability and operator independence. Reviews of the methods used should be undertaken at predetermined intervals or at times appropriate to a developed need.

Documentation   29.26   Laboratory documentation should be in line with the general guidance given in Chapter 9 on Documentation. When electronic or magnetic recording methods are used, see also Chapter 28 on Electronic Data Processing and Control Systems.

29.27   Retention samples should be regarded as part of the laboratory records.

29.28   It is useful to record test results in a manner that will facilitate comparative reviews of those results and the detection of trends. To assist this process, 'commodity files' may be established.

Records of Analysis   29.29   Details to be recorded on the receipt and testing of starting materials, packaging materials and intermediate, bulk and finished products are indicated in Chapter 9 on Documentation. Analytical records should contain:

(a) name of product or material and code reference;
(b) date of receipt and sampling;
(c) source of product or material (including supplier and country of origin);
(d) date of testing;
(e) batch or lot number;
(f) indication of tests performed;
(g) reference to the methods used;
(h) results;
(i) decision regarding release, rejection or other status; and
(j) signature or initials of the analyst, and signature of the person taking the above decision.

29.30   In addition to the above records, analysts' laboratory records should also be retained with the basic data and calculations from which test results were derived (e.g. weighing, readings, recorder charts).

Specifications   29.31   Specifications approved by quality control and including analytical parameters should be established for all raw materials, bulk, intermediate and finished products (see Chapter 9).

Testing   29.32   The persons responsible for laboratory management should ensure that suitable test methods, validated in the context of available facilities and equipment, are adopted or developed.

29.33   Samples should be tested in accordance with the test methods referred to, or detailed, in the relevant specifications. The validity of the results obtained should be checked (and as necessary, any calculations checked) before the material is released or rejected.

29.34   In-process control work carried out by production staff should proceed in accordance with methods approved by the person responsible for quality control.

Contract Analysis   29.35   Although analysis and testing may be undertaken by a contract analyst, the responsibility for quality control cannot be delegated to him/her.

29.36   The nature and extent of any contract analysis to be undertaken should be agreed and clearly defined in writing, and procedures for taking samples should be set out as discussed in 29.21–29.24.

29.37   The contract analyst should be supplied with full details of the test method(s) relevant to the material under examination. These will need to be confirmed as suitable for use in the context of the contract laboratory.

29.38   Formal arrangements should be made for the retention of samples and of records of test results. Protocols must be in place for timely reporting of the results that identify product non-conformance for product safety, legality and quality to the quality control manager or designate.

Accreditation   29.39   Where control laboratories are located within factories that are applying for certification to BS EN ISO 9001 or 9002, the laboratory is likely to be considered as part of the quality control system assessment. Adherence to good control laboratory practice should meet the technical standards demanded, although the documentation of the system will need to conform to the needs of the assessors. In the UK, many food laboratories will wish to seek accreditation under the UK Accreditation Service (UKAS) accreditation scheme. This scheme is appropriate to those laboratories that need the credibility of a public analyst or good contract laboratory. Proficiency testing schemes are run by the Central Science Laboratory, an executive agency of the UK Government Department for Environment, Food and Rural

Affairs (DEFRA). Participating laboratories will be asked to perform food chemistry [Food Analysis Performance Assessment Scheme (FAPAS)] or food microbiology [Food Examination Performance Assessment Scheme (FEPAS)] testing at regular intervals (see http://www.fapas.com for further details). The results will be published and approved laboratories listed in EU publications. Approval may be withdrawn from any laboratory if standards in any of the accreditation schemes are not maintained. Reference should be made to ISO Guide 25, which is the international equivalent of the UKAS system. Third-party laboratories used for testing should be accredited by a competent national authority to ISO 17025 or equivalent standard.

30   ENVIRONMENTAL ISSUES

Principle

In the Introduction, the intention was made clear to limit the Guide to matters having a direct bearing on the scientific, technological and organisational aspects affecting quality, legality and safety of products. For this reason, detailed consideration has not been given to the impact of the factory and its operations on the external environment. It is, however, acknowledged here that the management of any food manufacturing operation has general responsibility and, in most countries, legal obligations (with which it must be familiar) for these aspects.

General

30.1   The premises, equipment, personnel, manufacturing operations, intake of materials, despatch of products and treatment/disposal of unwanted by-products (waste materials, by-products unsuitable for human consumption, effluent, emissions of smoke, gases, fumes, dust, noise, light and odours that are offensive or that may cause taint elsewhere) must be controlled and comply with:

(a) all the appropriate environmental legislation in the country concerned;

(b) any specific legislation in the country in which manufacture takes place, relating to the treatment and handling of specific by-products; and

(c) the environmental legislation in the country in which manufacture takes place as well as any additional requirements of the local authority, and should have full regard to the responsibility of industry to take whatever steps are necessary to minimise, and preferably eliminate, any adverse environmental impact of its operations.

30.2   When building food manufacturing premises, consideration should be given to the local environment and the actions that may need to be taken to protect the environment. This is usually addressed by undertaking an environmental risk assessment that once developed and implemented must be routinely reviewed and updated as activities and practices on the site change. The factors to be considered include, but are not limited to:

(a) local activities that could impact on food production;

(b) chemical storage on-site and chemical mixing areas;

(c) fuel and oil storage on-site;

(d) potential for existing ground contamination or pollution from spillages;

(e) design of drainage systems including storm water and designated waste drains;

(f) sources of and conservation of water and energy;

(g) waste water treatment and the need for effluent treatment;

(h) storage areas for packaging awaiting recycling that minimise harbourage for pests;

(i) storage of other wastes awaiting collection, for example, in a poultry meat factory, the storage of blood, sludge or feathers.

30.3   Formal policies and protocols such as environmental policies, waste management and waste minimisation protocols (including recycling policies), dust, odour and noise management plans need to be developed, implemented and periodically reviewed to ensure that they are still valid, are being complied with and are effective. Resource management procedures should also be developed to effectively manage resources utilised within the manufacturing unit, for example, water, energy and materials.

30.4   Manufacturing organisations may seek third-party certification of their environmental management systems for compliance with standards such as EN ISO 14000:2004.

IPPC   30.5   **Integrated Pollution Prevention and Control (IPPC)** is operated under **the Pollution Prevention and Control (England and Wales) Regulations 2000**, and similar regulations for **Scotland and Northern Ireland** made under the **Pollution Prevention and Control Act 1999**, which implements the **European Community (EC) Directive 96/61/EC on IPPC**. The IPPC Directive has been codified **(Directive 2008/1/EC)**.

The IPPC Directive (96/61/EC) was implemented in England and Wales through the **Pollution Prevention and Control (England and Wales) Regulations 2000**, which have been replaced by the **Environmental Permitting Regulations 2007**.

The scope of the legislation includes all 'installations' treating and processing materials intended for the production of food from:

- animal raw materials (other than milk) at a plant with a finished product production capacity greater than 75 tonnes per day;
- vegetable raw materials at a plant with a finished product production capacity greater than 300 tonnes per day (average value on a quarterly basis);

Activities in this sector include production and preserving of meat, fish, potatoes; manufacture of fruit and vegetable juice; fruit and vegetable processing; milk processing; cereal processing; animal feed; pet food; bread, cakes and biscuits; sugar; chocolate and confectionery; pasta products; tea and coffee; and beverages and brewing.

- installations treating and processing milk, the quantity of milk received being greater than 200 tonnes per day (average value on an annual basis); and
- slaughterhouses with carcass production capacity of greater than 50 tonnes per day.

IPPC legislation requires manufacturing sites to identify their impact on air, soil and water using an integrated approach in order to develop environmental management systems to minimise the impact of the site. This could include the development of documented management systems to meet the needs of EN ISO 14000:2004. Consideration should also be given to the potential for activities and practices undertaken at the manufacturing site impacting on local environments and neighbours especially those sites that are environmentally sensitive.

Principle

In the Introduction, the intention was made clear to limit the Guide to matters having a direct bearing on the scientific, technological and organisational aspects affecting quality and safety of products. For this reason, detailed consideration has not been given to the safety and welfare of operatives (except insofar as their health and personal hygiene bears directly on the quality and safety of products). It is, however, acknowledged here that the management of any food manufacturing operation has general responsibility and, in most countries, legal obligations (with which it must be familiar) for the health and safety of its employees.

General

31.1   In the UK, **The Health and Safety at Work Act 1974** applies not only to employees but also to members of the public including customers, visitors and contractors. A duty is placed on employers with more than five employees to develop and implement a written health and safety policy.

31.2   In the UK, **The Control of Substances Hazardous to Health (COSHH) Regulations 1999, amended in 2004,** introduces a requirement for employers to carry out a formal assessment of all work that is liable to expose employees to hazardous substances, including liquids, solids, vapours, dusts and microorganisms. The assessment must evaluate the risks to health and the actions that are required to minimise that risk.

31.3   Other UK legislation that is applicable to food manufacturing premises includes:

(a) **The Electricity at Work Regulations 1989;**
(b) **The Health and Safety (Display Screen Equipment) Regulations 1992;**
(c) **The Manual Handling Operations Regulations 1992;**
(d) **The Personal Protective Equipment at Work Regulations 1992;**
(e) **The Workplace (Health, Safety and Welfare) Regulations 1992;**
(f) **The Reporting of Injuries, Diseases and Dangerous Occurrences Regulations 1995;**
(g) **The Provision and Use of Work Equipment Regulations 1998;**
(h) **The Work at Height Regulations 2005;**
(i) **The Control of Noise at Work Regulations 2005;**
(j) **The Control of Vibration at Work Regulations 2005;**
(k) **The Management of Health and Safety at Work (Amendment) Regulations 2006.**

Food & Drink – Good Manufacturing Practice: A Guide to its Responsible Management, Sixth Edition.
The Institute of Food Science & Technology Trust Fund.

PART II – SUPPLEMENTARY GUIDANCE ON SOME SPECIFIC PRODUCTION CATEGORIES

This part of the Guide is not intended to be a comprehensive commentary covering every product group in detail; it expands on general principles where it has been felt that there is a specific need to do so.

Food & Drink – Good Manufacturing Practice: A Guide to its Responsible Management, Sixth Edition.
The Institute of Food Science & Technology Trust Fund.
© 2013 John Wiley & Sons, Ltd. Published 2013 by John Wiley & Sons, Ltd.

32   HEAT-PRESERVED FOODS

General   32.1   Heat treatment is a method of preservation that will prevent or delay product deterioration or spoilage and inhibit the growth of pathogenic organisms. There are several methods of heat treatment including pasteurisation, ultra-heat treatment and sterilisation.

Heat Treatment   32.2   All such foods that are designed to be microbiologically stable must be heat treated to an extent that will prevent the growth of microorganisms under the storage conditions, including packaging, and during the period in which the foods are intended to be stored.

Commercial Sterility 32.3   Commercially sterile food is food that has undergone a heat treatment that will destroy all vegetative pathogens and organisms capable of growth or causing spoilage in the food under standard storage conditions, for example, low-acid canned food or aseptically packed product.

                     32.4   All low-acid foods having in any part of them a pH value of 4.5 or above and intended for storage under non-refrigerated conditions must be subjected to the minimum botulinum process, that is, one that will reduce the probability of survival of *Clostridium botulinum* spores by at least 12 decimal reductions, unless the formulation or water activity, or both, of the food is such that it can be demonstrated that growth of strains or forms of the organism cannot occur. The scheduled heat process required to achieve commercial sterility will be in excess of the minimum botulinum process ($F_o = 3$), as many organisms associated with the spoilage and economic loss of heat-preserved foods have greater heat resistance than *C. botulinum*. In addition to achieving commercial sterility, the scheduled heat process may be further extended for specific reasons, for example, to soften the bones in canned fish or to tenderise or texturise meats.

                     32.5   Low-acid foods packed in acidic carrying fluids or whose natural pH value has been otherwise lowered by the controlled addition of acids, should be subjected to at least a minimum botulinum process unless it can be established that the equilibrium pH, including the pH value at the cores of particles, is less than 4.2 within 4 hours of the end of the thermal process, thus providing a safety margin for non-uniformity of acidification.

                     32.6   The scheduled heat process is the key document for the manufacture of all heat-processed foods. In the UK, it is addressed by a number of guidelines available from Campden BRI.[1]

[1]http://www.campden.co.uk/publications/pubs.php.

Food & Drink – Good Manufacturing Practice: A Guide to its Responsible Management, Sixth Edition.
The Institute of Food Science & Technology Trust Fund.
© 2013 John Wiley & Sons, Ltd. Published 2013 by John Wiley & Sons, Ltd.

32.7 **There are UK legal definitions in Statutory Instruments of minimum scheduled heat treatments for milk, semi-skimmed and skimmed milk, milk-based drinks and cream in the Dairy Products (Hygiene) Regulations 1995, SI 1995 No 1086, as amended in 1996; for ice cream in SI 1959 No 734 as amended in 1962, 1963, 1982, 1985, 1990 and 1995; and for liquid egg, in the Egg Products Regulations 1993, SI 1993 No 1520.**

32.8 For efficient heat treatments, the physical, chemical and microbiological characteristics of each specific product need to be identified since these are critical factors that can affect the rate of heat penetration or heat distribution. These, therefore, need to be taken into account for plant design and processing regimes. Products should be defined in terms of their densities, rheological properties, water activity (a_w), pH, temperature, thermal properties, pressures, specific heats, microbial loading, gas content and corrosive properties. Electrical properties are critical when using electrical energy for sterilisation. The processing parameters set must be validated to demonstrate that they are capable of delivering safe food (see 32.26 and Chapter 3).

32.9 Discrete particles should be defined additionally in terms of size, shape, concentration and swelling or shrinkage. Dry ingredients should be thoroughly dispersed and wetted. Noncondensable gases should be disengaged.

32.10 Scheduled heat processes are derived by measuring the rate of heat penetration into the product and integrating the time/temperature exposure, hence determining the total lethal heat that the product receives within a particular process regime. For low-acid foods, or foods otherwise designated as requiring at least a minimal botulinum process, 'worst-case' assumptions involving the lowest temperature reached and shortest retention time at it, should be made for *C. botulinum* spore destruction kinetics (see 32.4).

32.11 All new products or changes in the manufacturing operations, formulation and, if appropriate, containers for existing packs should be fully evaluated as to their effect on the rate of heat transfer through the product before commercial production is undertaken. Effective validation must be undertaken to ensure that such modifications, and the associated changes to processing conditions that are required, are effective. This activity must be recorded and records maintained.

32.12 Products not given at least a minimum botulinum process, or are not otherwise shelf stable at ambient temperature, should carry clear instructions on storage conditions, including maximum storage temperatures.

32.13 All containers should be indelibly marked with a code indicating at least the place and date of production. Further information on the time of production and the production line can be very valuable. Empty containers or reel stock should be checked on receipt to ensure that they comply with the agreed recommendations for the product. They should be stored and handled so as to prevent their becoming contaminated or damaged and the integrity of the container thereby affected. Containers conveyed into a filling machine should be clean and flawless.

32.14   All batch and continuous processing equipment should be fitted with direct reading (indicating) temperature probes and automatic time and temperature recording instruments. These should be calibrated at designated intervals (see Chapter 25). Records of calibration and processing records, including data from temperature monitoring and automatic recording devices, should be kept for at least 3 years from the date of production depending on the duration of the product or customer requirement. Monitoring may be undertaken by on-line or quality control personnel. Personnel undertaking monitoring need to be aware of the critical nature of such monitoring and the actions to be taken in the instance that there is a product or process failure identified at a product safety critical control point (see Chapter 3).

32.15   Food processing equipment should be internally clean and disinfected, correctly assembled, free from 'dead legs' and with control systems that have instruments of reproducible accuracy that are routinely calibrated. Procedures must be in place to address the actions to be taken in the event of equipment calibration failure (see Chapter 25).

32.16   Water used for manufacturing purposes, including that used in making up products or likely to come into direct contact with the product, must be of potable quality (see Chapter 13) and free from any:

(a) substances in quantities likely to cause harm to health;
(b) substance at levels capable of causing accelerated internal corrosion of metallic containers and closures or causing taints; and
(c) harmful microorganisms.

To maintain the required microbiological standards, the water should, if necessary, be chlorinated or otherwise adequately treated (see 13.2 and 32.32).

32.17   Mechanical unloading and handling systems should be designed so that:

(a) driers can be located as early as possible in the conveying system;
(b) there is minimum contact between container closures and the conveyor surfaces;
(c) abuse of containers is kept to a minimum; and
(d) conveyor surfaces may be easily cleaned and disinfected as frequently as is necessary to maintain proper standards of hygiene (see Chapter 14).

32.18 Regular microbiological surveillance including swab tests should be made on conveyors and other contact surfaces to establish the effectiveness of sanitation programmes (see 14.13).

32.19 To reduce the potential for cross-contamination between high-risk and low-risk areas (HRAs and LRAs), the post-process area and its personnel should be segregated from other sections, particularly from the preparation area of the plant. In the instance of manufacturing units that process raw and cooked meats, strict segregation of personnel should be undertaken including segregation of welfare facilities, designated staff, designated clothing, tools and equipment (see 33.25).

32.20 Processed containers should be loaded into clean dry cartons or other suitable outer packaging material and stored in areas specifically allocated for this purpose. Care should be taken to minimise any damage to processed containers that could compromise seal integrity.

32.21 Samples of finished packs may be incubated as part of the overall quality control programme. However, the acceptability of the manufacturing operations should not be judged solely on the results of such tests, but rather that in most situations they are forms of verification and provide cumulative and retrospective confirmation of the efficacy of the process control and hygiene operations.

32.22 It is essential to ensure product integrity after processing and handling. Suitable precautions should be taken to ensure that package structures, including seals and seams, are free from defects that would otherwise impair their effectiveness as microbiological barriers, and that the food does not cause breakdown of the packaging materials with which it is in contact.

32.23 The products should be suitably protected for handling and stacking in storage. They should bear identification marks that are traceable to the production records derived from the manufacturing process (see Chapter 10).

32.24 Transport may present risks to package integrity caused, for example, by tilt and vibration, in respect of which correct handling procedures should be adopted (see Chapter 23).

32.25 Every effort should be made to take into account factors that may affect the continuing integrity of the products after they leave the immediate control of the producer. This, for example, includes handling during transportation, in wholesale and retail outlets and by the consumer. Labels warning against the use of case hooks and careless use of case opening knives should be adopted.

In-Container Heat Treatment

32.26 The scheduled heat process should take into account all critical factors that may affect the rate of heat transfer in the containers. It should be established by competent and properly trained personnel using accepted scientific methods. The scheduled heat process should be validated during the product and process design and re-validated in the event of changes or amendments to the process or product design. Records giving full details of how the scheduled heat process was established and confirmed as appropriate through validation should be retained permanently on file.

32.27 Product preparation and filling operations, for example, setting of headspace that may affect the integrity of the container closure or the rate of heat penetration should be carefully controlled.

32.28 The efficiency of the closing or sealing operation should be checked before processing begins and kept under constant control to ensure the integrity of the container closure or seal. Records of quality control assessments should be kept for at least 3 years from the date of production depending on the duration of the product and customer requirements.

32.29 Washing of filled and closed containers, should, if required, be undertaken before the heat process is given.

32.30 Temperature distribution tests should be undertaken on all processing equipment and a venting schedule established for each type of processing unit. All changes in services and their layout to processing equipment or alterations in the method of loading batch retorts should be fully evaluated for their effect on the temperature distribution and adequacy of venting schedule.

32.31 For batch retorts, there should be a method of checking that no batch of containers has bypassed the heat process.

32.32 Potable water including water for cooling containers should be free from harmful microorganisms. Aerobic plate counts (APCs) carried out using nationally accepted methodology should not exceed 100 organisms/mL after incubation for 5 days at 25°C or a recognised equivalent thereto. Alternatively, deep borehole water run directly to retorts, or municipal mains water may be used, provided that an APC as indicated above is consistently achieved. All water used for cooling should be sufficiently chlorinated with an adequate contact time (for 20 minutes or more, depending on the pH) before use such that free residual chlorine (FRC) can be detected in the water after it has cooled the containers (see 13.2). Tests for FRC should be undertaken at a frequency derived by risk assessment, ideally through continuous monitoring. It is not common practice to use other chemical disinfectants to sanitise water for cooling purposes, but where approved materials are contemplated for use they should achieve the same APC quality standard and have a residual present in active form after the cooling cycle.

32.33 Cooled containers should not be manually handled while still wet.

Heat Treatment
Followed by
Aseptic Packaging

32.34 Aseptic technology is more complex than in-container sterilisation and needs to be thoroughly understood by the processor. It involves the production of a commercially sterile hermetically sealed package of food by pre-sterilising the food, followed by cooling, filling and sealing it in sterile containers under scheduled processes, which must be applied in order to achieve and maintain commercial sterility. Assurance of attaining these standards comes from in-process controls. Retrospective quality control analysis can only indicate the success or failure to maintain the required standards. The microbiological objective is to achieve and maintain commercial sterility.

32.35 Pre-production sterilisation of all food contact surfaces should take into account any location that may be slow to attain the set temperature. Non-food contact surfaces may be chemically treated to achieve sterilisation. Chemicals that may be used include hydrogen peroxide. The integrity of the total system should be checked at this stage.

32.36 Aseptic fillers should be cleanable, sterilisable and capable of being isolated by a separating sterile barrier and with control systems that have instruments of reproducible accuracy, which are routinely calibrated. Aseptic filling machines may be affected in their performance by dust or soil in the external environment, and therefore careful consideration should be given to their location. A filling environment should have internal surfaces suitable for cleaning and microbiological sterilisation. Any inflow of sterile air or gas should have suitable flow, velocity and direction. With laminate packaging, the operation of the sealing equipment is also critical in ensuring that the packaging is adequately sealed to prevent air or bacterial ingress. Quality control personnel and/or machine operators should ensure that the machines have been set up correctly and the resultant sealing is within specification before production commences.

32.37 Aseptic fillers should be appropriate for both product and packaging. Product temperature should also be compatible with the packaging material used. Control over dispense volume is critical since this may affect the integrity of the aseptic seal of the packages, or the internal stress caused by the product in the finished packages.

32.38 During start-up, standby or shutdown of an aseptic filling machine, special conditions may apply, for example, in the supply of decontaminated air or gas, application of heat or application of chemicals. Care must be taken to ensure that all personnel health and safety procedures are followed and any fire risk is minimised.

32.39 Pre-packing sterilisation of a product needs to take into account the critical factors to be considered when determining a scheduled heat process (32.8–32.11).

32.40 In continuous-flow sterilisation, the control system should ensure that the correct product sterilisation temperature and holding time are achieved. It should be designed to fail-safe in the event of malfunction, and underprocessed food should be routed out of the system and not transferred to the aseptic storage or filling section. The plate packs on pasteurisers should be routinely checked for integrity and to ensure that there are no pinholes in the plates. The relative pressure of the product and the cooling water should be reviewed to ensure that the risk of cooling water contamination through a pinhole/crack is minimised.

32.41 Direct heating systems may use electrical energy instead of steam. The energy and heat transfer mechanisms are different and should be understood by those responsible for the process design, the control system and its operation.

32.42 During continuous-flow product cooling, there should be an absence of microbiologically contaminating leaks.

32.43 Equipment used for batch sterilisation of product should have sensors, indicators, controllers and recorders for temperature and pressure. Bulk mixing patterns are important where they relate to temperature distribution. There should be a capability of aseptic transfer of product and an absence of microbiologically contaminating leaks.

32.44 Any methods used for microbiological decontamination of packaging should be compatible with the packaging materials and also take into account the microbiological loads. The methods should be designed using proven procedures. When scheduled processes for microbiological sterilisation of packaging materials depend on synergistic effects, these are critical factors that should then be monitored.

32.45 Heat-based processes depend on suitable temperature, pressure, humidity, time of exposure and extent of air venting, and all processes should be formally validated and then re-validated at routine intervals. Monitoring and verification activities should be adopted to ensure that critical limits as well as target levels and tolerances have been complied with during processing.

32.46 Chemical applications depend on type, concentration, dose, temperature, coverage and contact time, and all processes should be formally validated and then re-validated at routine intervals. Monitoring and verification activities should be adopted to ensure that critical limits as well as target levels and tolerances have been complied with during processing.

32.47 Ultraviolet radiation has biocidal activity that is dependent on wave and band frequency, power, distance, reflectance, temperature, exposure time and age of lamp. Dust particles may have a shielding effect. Ionising radiations exert sterilising effects depending on the dose. Therefore applications of this nature must be validated and then re-validated at routine intervals. Monitoring and verification activities should be adopted to ensure that critical limits as well as target levels and tolerances have been complied with during processing.

32.48 Packaging materials that are microbiologically decontaminated by the foregoing methods should be protected from recontamination.

32.49 Packaging materials may require special conditions of storage and must be stored and handled to minimise damage that can affect packaging integrity especially with laminate packaging.

32.50 Temperatures and humidity should be controlled within acceptable limits, and additional safeguards should be applied where pre-sterilised packaging materials or packages are used.

33   CHILLED FOODS

General   33.1   Chilled foods are perishable foods that, in order to extend the time during which they remain wholesome and safe to eat, are kept within controlled and specified conditions normally below 8°C. Chilled foods may be divided into those that potentially offer a low risk of the growth of pathogenic organisms and those where the risk is theoretically high. The factors that invoke the need for additional care in the manufacture of chilled foods are the:

(a) perishability of raw materials;
(b) minimal processing to maximise sensory quality;
(c) potential for spoilage and/or pathogen growth;
(d) rapid despatch of finished products; and
(e) chilled food chain requirements.

33.2   Given the short shelf life of chilled prepared foods, the emphasis must be on the use of hazard analysis critical control point (HACCP) to develop a system to control processes, premises, personnel and the hygienic status of ingredients/raw materials used, rather than end-product testing (see Chapter 3). Product safety must be determined by the proper consideration of the following:

1. ingredient hygienic quality;
2. product formulation/characteristics;
3. processing parameters;
4. allergen control;
5. intended use of product;
6. storage and distribution conditions;
7. manufacturing hygiene; and
8. shelf life.

Ingredients   33.3   Points to consider:

• Which pathogen(s) and level of contamination might be present?
• Is there a possibility of preformed toxins?
• What are reasonable specification levels to apply to minimise risk?
• What further processing is to be applied?
• Does the shelf life of the ingredients exceed the shelf life of the finished product?

33.4   Specifications for ingredients/raw materials should include microbiological standards. Ideally, on-site facilities for microbiological testing should be available. Where this is not available, provision should be made for adequate microbiological testing at a laboratory that has

Food & Drink – Good Manufacturing Practice: A Guide to its Responsible Management, Sixth Edition.
The Institute of Food Science & Technology Trust Fund.
© 2013 John Wiley & Sons, Ltd. Published 2013 by John Wiley & Sons, Ltd.

suitable third-party accreditation (see Chapter 29). The UK Health Protection Agency (HPA) has produced guidelines for assessing the microbiological safety of ready-to-eat foods placed on the market.[1]

33.5 Perishable ingredients/raw materials should be purchased only from approved suppliers who should furnish regular test results and agree to warn the purchaser of any problems in maintaining the standards agreed (see Chapter 16).

33.6 Deliveries of highly perishable raw materials should not be accepted if their temperatures fall outside agreed specified ranges. Guidance on recommended storage conditions should be given on outer packaging. Temperature checks undertaken must be recorded and the equipment used must be calibrated at designated intervals.

33.7 Inspection of perishable raw materials should be based on risk assessment and may include rapid indicative testing methods. However since testing in itself is not a control measure as recognised by HACCP, the emphasis is on proactive controls.

Product Formulation/ 33.8 The growth of pathogenic microorganisms can be controlled by
Characteristics product formulation/characteristics. This might include:

- adjustment of pH; and/or
- adjustment of water activity (a_w); and/or
- addition of chemical preservative.

An individual factor, such as pH, may be used to reduce microbiological growth. Introducing an additional factor, such as water activity, may produce a synergistic effect; that is, the combination of the two factors as hurdles reduces microbiological growth to a greater than expected or calculated extent. Relatively small changes to both hurdles together (e.g. pH + a_w) may be as effective as large changes to either hurdle in isolation. Chemical preservatives are rarely added direct to chilled prepared foods, but may be used in the preservation of chilled cooked meats as an example. However, where they are used, this must be in compliance with the regulatory requirements of the country concerned, and should be at the minimum effective level to ensure product safety, while not themselves presenting a food safety hazard.

Processing 33.9 This aspect should be considered under the following subdivisions
Parameters and outcomes:

(a) *Heat Treatments*
- none or less than 70°C for 2 minutes. Possibility that all pathogens present will survive.
- heated to 70°C for 2 minutes (or equivalent). All vegetative pathogens present will be reduced to an acceptable level

---

[1]http://www.hpa.org.uk/webc/HPAwebFile/HPAweb_C/1259151921557.

(6 log reduction), for example, *Listeria monocytogenes*, *Staphylococcus aureus*, Salmonellae and verocytotoxigenic *Escherichia coli*. However, spores and preformed toxins may persist. The potential for recontamination by vegetative pathogens must be limited by the use of a high-risk area (HRA) (see 33.17).

- heated to 90°C for 10 minutes (or equivalent). In addition to vegetative pathogens, spores of psychrotrophic (non-proteolytic) *Clostridium botulinum* will be reduced to an acceptable level (6 log reduction). However, more heat-resistant spores, for example, strains of *Bacillus cereus* and some preformed toxins may persist. Measures to prevent the outgrowth of psychrotrophic *C. botulinum* must be in place where chilled products have a shelf life of more than 10 days.

(b) *Cooling*
- heated product should be cooled as quickly as possible through the so-called danger zone that is the temperature range of 63°C to 5°C. This will minimise risk of spore germination and outgrowth. The time taken for cooling will vary from product to product but, as a guideline, should be no more than 2 hours. Blast chillers are often used to achieve this temperature profile.

(c) *Packaging*
This should be considered under the following subdivisions and outcomes:
- product cooked in-pack. Pathogens will be eliminated to the extent indicated under 'Heat Treatments' (see above), and there will be no opportunity for recontamination that may present a food safety hazard (assuming complete integrity of the pack/seal).
- product cooked, cooled and assembled. Pathogens will be eliminated to the extent indicated under 'Heat Treatments' (see above), but there is risk of recontamination during assembly that may present a food safety hazard.

Modified atmosphere packaging (MAP) or vacuum packaging may be used to reduce microbiological growth in conjunction with chilled storage but will not necessarily inhibit the growth of pathogens. There may be particular concerns with respect to *C. botulinum*. It is important that the effectiveness of MAP or vacuum packaging is assessed in each case, with reference as necessary to specialist advice.

Intended Use of the Product

33.10 The key point to consider is whether the product is to be eaten:

- without further heating (i.e. is ready to eat);
- following domestic reheating (i.e. requires reheating only prior to consumption); or
- following domestic cooking (i.e. requires a heat process equivalent to at least 70°C for 2 minutes prior to consumption).

On-pack instructions are an important control measure to ensure that correct procedures are followed and food safety risk is reduced. Any instructions provide by the manufacturer must be validated to demonstrate their efficacy with records retained, and in the event of a change to the product, the instructions must be revalidated, for example, increasing the size of chicken pieces in a ready meal or pie.

The Chill Chain 33.11 Temperature and time control are the principal controlling factors for the safety of chilled foods. Effective temperature control throughout the chill chain is particularly important to slow or inhibit the growth of pathogenic bacteria. Chilled foods, for reasons of safety or quality, are designed to be stored at refrigeration temperatures (at or below 8°C, targeting 5°C) throughout their entire life. The performance of the proposed distribution chain should be validated and monitored by the responsible party and taken into account when specifying shelf life. The number and location of temperature probes in storage should be such as to ensure effective monitoring. The manufacturer must consider whether automatic failure alert systems such as visual or audible alarms or processing fail-safe systems should be in place. If such systems are used, then they must be monitored at a designated frequency by quality control or production personnel who have been adequately trained and understand the significance of such failures and the appropriate action to take.

Records should be maintained of the monitoring and verification activity and the corrective action taken in the event of failure. If these points in the process are deemed within the HACCP plan to be critical control points (CCPs), there must be an appropriate level of awareness among all personnel of the consequence of a target level and tolerance and/or a critical limit being exceeded. Visual or audible alarms can be designed into the process to be activated if a critical limit or target level is exceeded, for example, at the metal detection stage and diversion of product at a pasteuriser. The automatic failure alert systems must be tested on a routine basis and records maintained of the tests and the results and any remedial action taken. Calibration procedures must be in place for all automatic failure alert systems (see Chapter 25).

Manufacturing Hygiene 33.12 The purpose of establishing specified standards of personnel and premises hygiene is to control the hazard of microbiological contamination and cross-contamination. The level of hygiene required will depend on the risk of and/or the consequences of cross-contamination (see Chapters 11, 12, 14 and 15).

33.13 Chilled foods are manufactured using a wide variety of raw materials, processes and packaging systems and therefore must be expected to have both differing microbiological profiles after manufacture or storage and differing shelf lives, but must be microbiologically safe at the point of consumption. Thus products made entirely from ingredients that are heated in a container or are assembled from heated ingredient(s) under the special hygiene conditions defined as 'high risk', are designed to be free from vegetative pathogens, but not all spore formers. Those containing raw ingredients will from time to time contain vegetative pathogens such as *Listeria* as well as toxin producers and/or spore formers. This difference must be taken into account when specifying shelf life in terms of time and temperature and consumer instructions. Since *Listeria* is able to grow under chilled temperature conditions, this is an important organism to control. In doing so, the risk of other vegetative pathogens being present is also minimised.

Shelf Life

33.14 Shelf life depends on the control of all the preceding factors, but must be validated by challenge testing and/or modelling for each product and process for defined chill storage conditions. It must be recognised that the integrity of the whole of the chill chain is vital to ensure the safety and quality of chilled foods.

33.15 By developing a product utilising a combination of factors such as raw material quality, hygienic processing, temperature, water activity, acidity and modified atmosphere, microbiological growth can be controlled and thus spoilage and/or food-borne illness prevented. The choice and combination of hurdles will determine the shelf life and the conditions of use of the products.

33.16 Consult CCFRA Guideline G46 'Evaluation of Product Shelf Life for Chilled Foods' (2004) and 'Shelf life of ready to eat food in relation to *L. monocytogenes*—Guidance for food business operators' (CFA/BRC/FSA, 2010) for further information on the determination of shelf life. Further guidance can be found in the guidance section of this publication (Appendix IV).

Application of HACCP

33.17 There are distinct terms that need to be considered when applying HACCP to the safety of chilled foods, and they are described below:

Ready to Eat (RTE): Food intended by the producer or the manufacturer for direct human consumption **without the need for cooking or other processing** effective to reduce to an acceptable level or eliminate microorganisms of concern (i.e. cold eating).

Ready to Cook (RTC): Food designed to be given a heat process by the consumer that will deliver a 6-log kill with respect to vegetative pathogens (a minimum process equivalent to 70°C for 2 minutes) throughout all components.

Ready to Reheat (RTRH): Food manufactured in a high-care area (HCA) or HRA that has been designed to be reheated by the final consumer.

Low-Risk Area (LRA): An area where good manufacturing practice (GMP) standards are in place as described within this publication, but the area and the practices have not been specifically designed to minimise microbial contamination, for example, raw material intake, storage areas of RTC foods and packaged product where the product is fully enclosed (see 32.19).

High-Care Area (HCA): An area designed to a high standard of facility specification and hygienic design where practices relating to personnel, ingredients, equipment and environment are managed to **minimise** microbial contamination of a RTE or RTRH product containing uncooked ingredients. HCA include:

- areas where RTE and RTRH food is being produced/ assembled; and
- areas where RTE/RTRH ingredients **not** thermally processed (minimum 70°C for 2 minutes) **but** have been decontaminated **and** grown/produced to RTE standards are stored and handled.

High-Risk Area (HRA): An area designed to a high standard of facility specification and hygiene design where practices relating to personnel, ingredients, equipment and environment are managed to **minimise** microbial contamination of a RTE or RTRH product comprising only cooked ingredients:

- areas where RTE and RTRH food is being produced/ assembled; and
- areas where **only** thermally processed foods (minimum 70°C for 2 minutes for <10-day shelf life) are stored and handled.

Product safety requires the application of HACCP principles to the design of plant, product and process, the management of hygiene procedures and the control of the process. HACCP is an evolving risk management system – it must be reviewed when any change is made to plant, raw material specifications, process or packaging (see Chapter 3).

33.18 Figure 33.1 represents decision trees formulated by the Chilled Food Association (CFA) (http://www.chilledfood.org) whose permission to include them and assistance with the preparation of this chapter are gratefully acknowledged. The first decision tree shows the basic decision pathway. Both the heat treatment (if any) applied in manufacturing the food product, any microbiological hazards arising between the heat treatment and final pack sealing and the final status of the food (i.e. ready to eat, ready to reheat or ready to cook) must be considered.

In general, only cooked-in-pack products are free of any risk of recontamination between heat treatment and final pack sealing and can therefore be produced in a LRA. The completed pathway through the decision tree shows the **minimum** hygiene status of the product handling environment between heat treatment and final pack sealing.

33.19 A detailed description of the requirements in each of the above areas is given in the CFA's 'Best Practice Guidelines for the Production of Chilled Food', Fourth Edition (2006). The standards highlighted are **minima** for each type of product. Higher standards can be used, but due regard should be paid to the potential for cross-contamination between lower-status and higher-status products

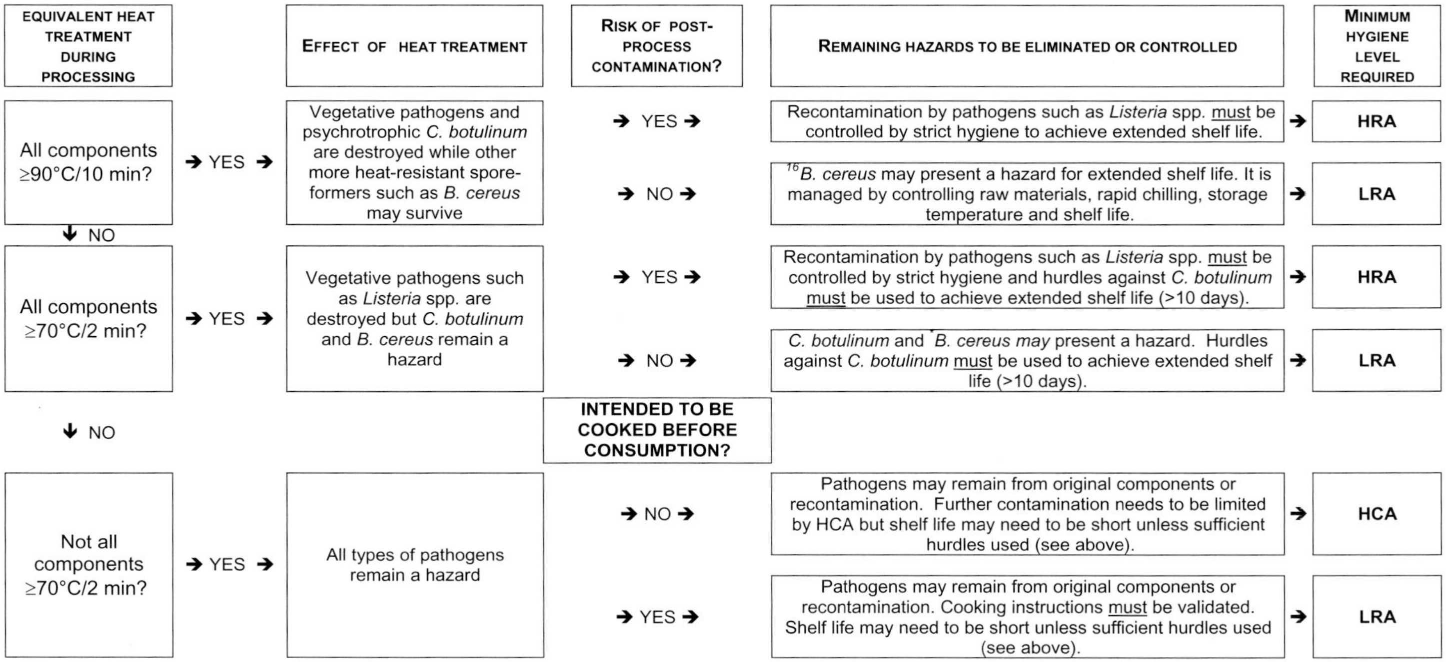

EQUIVALENT HEAT TREATMENT DURING PROCESSING · **EFFECT OF HEAT TREATMENT** · **RISK OF POST-PROCESS CONTAMINATION?** · **REMAINING HAZARDS TO BE ELIMINATED OR CONTROLLED** · **MINIMUM HYGIENE LEVEL REQUIRED**

All components ≥90°C/10 min? → YES

Vegetative pathogens and psychrotrophic C. botulinum are destroyed while other more heat-resistant spore-formers such as B. cereus may survive

→ YES → Recontamination by pathogens such as Listeria spp. must be controlled by strict hygiene to achieve extended shelf life. → HRA

→ NO → [16] B. cereus may present a hazard for extended shelf life. It is managed by controlling raw materials, rapid chilling, storage temperature and shelf life. → LRA

↓ NO

All components ≥70°C/2 min? → YES

Vegetative pathogens such as Listeria spp. are destroyed but C. botulinum and B. cereus remain a hazard

→ YES → Recontamination by pathogens such as Listeria spp. must be controlled by strict hygiene and hurdles against C. botulinum must be used to achieve extended shelf life (>10 days). → HRA

→ NO → C. botulinum and B. cereus may present a hazard. Hurdles against C. botulinum must be used to achieve extended shelf life (>10 days). → LRA

↓ NO

INTENDED TO BE COOKED BEFORE CONSUMPTION?

Not all components ≥70°C/2 min? → YES

All types of pathogens remain a hazard

→ NO → Pathogens may remain from original components or recontamination. Further contamination needs to be limited by HCA but shelf life may need to be short unless sufficient hurdles used (see above). → HCA

→ YES → Pathogens may remain from original components or recontamination. Cooking instructions must be validated. Shelf life may need to be short unless sufficient hurdles used (see above). → LRA

Figure 33.1. Decision tree to determine the minimum hygienic status required for chilled products.

within a single-status environment. Other products must not be produced in a HCA unless HACCP shows there are no additional risks to all products. Expert microbiological advice should be sought if necessary.

Quality Control 33.20 Microbiological testing of raw materials, intermediate and finished products should aim at monitoring and verifying standards and trends, not simply to accept or reject on the basis of results. In cases of unsatisfactory results, recall procedures would be necessary; therefore, any adverse trend in results calls for immediate investigation of process conditions and controls including in the supply chain, thus reducing the risk of a recall situation developing (see Chapter 19). It should be determined whether a positive or negative release system is in place with regard to microbiological testing.

33.21 Production programmes should provide sufficient time for adequate cooling; a deadline for acceptance of late orders should be established to ensure this. Care should be taken to avoid condensation on cold product by exposure to warm product or humid conditions.

33.22 The correct functioning of refrigerated despatch vehicles should be checked before loading, and product temperatures should be checked and marked on delivery notes before loading. At maximum load, despatch vehicles should be able to maintain specified temperatures and ensure that products carried remain within appropriate temperature profiles throughout the load and for the whole journey. A system should be in place to monitor the vehicle's operating temperature. This could include manual logging of data on the appropriate record during transport of the product, inspections using a calibrated temperature probe at each delivery point (where the temperature is formally recorded) or data logging devices where the data can be electronically downloaded. All equipment must be checked at routine intervals to verify that results obtained are accurate and the equipment is working within an acceptable tolerance. Documented transport breakdown procedures should be in place, which define the actions to be taken in the event of vehicle or refrigeration unit breakdown, and the records to be maintained that detail the incident and corrective action taken.

33.23 The quality control function should liaise with distribution management and customers to establish that there is adequate provision for maintenance of the temperature-controlled chain during transport and storage with the need for this being understood by all personnel. Major control points also include the occasions when products may not be in a temperature-controlled environment during the loading and unloading of vehicles, chilled stores and retail display cabinets.

33.24 Any frustrated deliveries that are returned should be examined by quality control, who should ensure that only goods in standard condition are accepted for further despatch. Other returns, suitably labelled, should be kept separately until disposal is agreed and actioned.

33.25   Special conditions must be provided for the production of high-risk products. Detailed requirements are set out in the CFA Guidelines. The production area should be completely separated from other areas from floor to ceiling. Positive pressure ventilation with micro-filtered air at the appropriate temperature and humidity should be provided. However, rather than a specific pressure being recommended, which is technically challenging to measure, a minimum number of air changes per hour is the preferred control parameter. Entry and exit should be through changing rooms, with 'no-touch' washing facilities. Emergency exits from the production area should be alarmed so that access via these exits can be monitored. Personnel must wear, as instructed, designated internal footwear and clothing. Only fully processed food components and packaging materials should be admitted, with entry through hatches/airlocks. Chill rooms for products awaiting pre-packing should be part of the designated special area. Construction and finish should provide for easy cleaning and disinfection. Only essential equipment and materials should be permitted in the production area. External packaging and ingredients entering the area may be subject to a sanitising procedure. Production should be interrupted for frequent thorough cleaning and disinfection. The frequency should be determined by risk assessment, which should be formally recorded and retained together with any results of verification activities such as product sampling or swab results. Otherwise cleaning and disinfection or sanitation may be undertaken between product types or varieties in addition to other hygiene activities, for example, between the assembly of different sandwich types.

33.26   Only named, authorised and medically cleared individuals should enter the production area. This includes management, technicians, engineers and visitors. It must be impressed on visitors, contractors and staff how critical it is to maintain personal hygiene and minimise the risk of product contamination (see Chapter 11). Medical screening should be undertaken with emphasis on:

(a) skin health—absence of acne, boils, other infections, wounds and burns;
(b) ear, nose and throat infections;
(c) gastrointestinal disorders; and
(d) contact with known cases of food poisoning.

33.27   Pre-employment screening should be made by a physician with experience in occupational medicine. The health status of selected employees should be reviewed by the physician at appropriate intervals, and a daily check, before starting work, should be made by an occupational nurse or designated individual.

33.28   Before entering the special area, outer garments should be changed for clean overalls and internal footwear. Effective head covering, coloured distinctively, should be worn. Beards and long hair should be discouraged, but if worn, should be fully contained. Hands should be washed with a non-aromatic antiseptic soap or lotion. Personnel should not use the same catering or toilet facilities as personnel from other areas (see Chapter 11).

33.29 Designated equipment should be provided in each area to prevent cross-contamination. This includes maintenance equipment that should be designated to specific areas, for example, HRA, so that the tools can only be used in that area.

34   FROZEN FOODS

General

34.1   Frozen foods are preserved by freezing and storing at temperatures cold enough to inhibit the growth of microorganisms and to retard chemical and physical reactions to a negligible rate. Primary long-term (3 months or more) stores should be maintained at or colder than −26°C. Temperature fluctuations of more than ±2°C should be avoided and the frequency of the variation kept to a minimum. The monitoring and continuous recording of storage air temperature using calibrated probes is essential. The number and location of temperature probes should be such as to ensure effective monitoring. Air temperatures in cold stores should also be manually checked and logged at least once every 24 hours. Warmer than specified warning and action limits should be specified and responsibilities for taking corrective action defined. Records should be maintained that detail any incident and the corrective action taken. Raw material and product temperatures should also be monitored to ensure that they remain within acceptable levels. Records should be retained of such activities. The temperature of raw materials and products should not be higher than −18°C. If the products have a temperature above −18°C, they should be quarantined, inspected, records maintained and appropriate action taken. Hazard analysis critical control point (HACCP) should be used as a management tool to develop a HACCP plan and associated food safety management system (FSMS) and quality management system (QMS). At critical control points (CCPs), critical limits and target levels and tolerances should be established so that a tendency towards a loss of control can be detected and rectified rather than waiting until a potentially unsafe product has been produced (see 34.11).

Personnel and Testing Facilities

34.2   Personnel engaged in management, production, quality control and maintenance should receive the necessary training and conform to medical and hygiene requirements described in Chapter 11. Production should be supervised by appropriately qualified personnel. Facilities should be available for microbiological and quality testing.

Raw Materials

34.3   Raw material specifications, acceptance and storage should be in accordance with the provisions stated in Chapters 2 and 6.

34.4   Raw materials should be selected for their ability to withstand freezing and thawing cycles, for example, using specially modified starches to prevent syneresis.

34.5   Incoming high-risk foods such as meat, poultry and fish should be subjected to microbiological testing. Such raw material should be temporarily quarantined until test results are available (see Chapter 6), that is, a positive release system must be in place.

Food & Drink – Good Manufacturing Practice: A Guide to its Responsible Management, Sixth Edition.
The Institute of Food Science & Technology Trust Fund.
© 2013 John Wiley & Sons, Ltd. Published 2013 by John Wiley & Sons, Ltd.

34.6 Although microbial load sometimes may be reduced by freezing, this cannot justify the use of microbiologically doubtful raw materials. Freezing does not destroy any microbial toxins that may have been formed, and does not eliminate the possibility of microbial problems at a later stage. Enzymes that have not been inactivated by treatment before freezing may continue to act after freezing.

Formulation

34.7 The freezing of food will not improve its quality. The flavours of some spices and other flavours can change quite drastically on freezing and subsequent storage. In products where enzyme systems are not fully inactivated, freeze concentration of enzymes and other substances can cause damage to texture and flavour. The sensory evaluation step of product development should be done after passing through one freeze/thaw cycle, with ultimate product life requirements being taken into account.

Processing

34.8 Preparation of raw material should be carried out in areas segregated from those concerned with cooked, blanched or finished product.

34.9 High-risk raw material, for example, poultry, meat or fish, should be prepared in rooms completely isolated, by walls from floor to ceiling, from cooked or finished products. There should be no interchange of staff or equipment between the areas. The preparation rooms should be maintained at a cool temperature and provided with air, filtered to remove microorganisms, at 8°C under positive pressure (see 33.25).

34.10 For the production area and equipment, a 'clean-as-you-go' regime should be instituted, debris and gross soiling being constantly removed, and each day a major strip-down and cleaning and disinfection operation should be carried out according to a documented cleaning schedule.

34.11 CCPs should be established at points in the process where microbial proliferation could occur if not suitably controlled. Monitoring and verification of those CCPs could involve:

(a) measurement of temperature and residence time of a particular sample at frequent intervals throughout the shift (monitoring if the product can be brought back under control; otherwise this is a verification activity);
(b) assessment of the degree of build-up of food debris or hold-up of food material at frequent intervals throughout the shift (monitoring if the product can be brought back under control; otherwise this is a verification activity);
(c) collection and microbiological testing of swabs or food samples at designated intervals during the production run. If the product is positively released on the basis of the sample results, this is monitoring; otherwise if a negative release system is in place, this is a verification activity.

It is critical that production staff understand whether the activities they are undertaking are monitoring or verification so that appropriate action can be taken if a failure is identified, which ultimately could be a product withdrawal or recall. Timeliness of decision making is fundamental to effective good manufacturing practice (GMP) being demonstrated.

Results of monitoring and verification should be recorded, preferably as graphs or histograms, and then assessed for signs of trends. Hygiene audits should also be analysed for trends especially if this includes further verification by swab analysis. From such records, the key performance indicator (KPI) or critical success factor (CSF) for a particular line can be established, and any adverse departure from these indicators is countered by implementing corrective action to improve hygiene. It should be borne in mind that if the product has to be thawed before cooking by the end user, high mould and yeast levels within the food can be a special problem. A quality plan should also be implemented to monitor quality attributes at key quality control points (QCPs) in the process.

34.12   Products or ingredients that are heated or cooked during manufacture should be cooled as quickly as possible to below 8°C.

34.13   Throughout the preparation stage, every effort should be made to keep the food cool and to keep it progressing along the production line.

Storage   34.14   Storage conditions and controls are addressed in Chapter 18. Frozen storage must be adequately controlled to prevent excessive build-up of ice on walls, floors and ceilings. Consideration should be given to personnel health and safety when working in these areas including the provision of alarms if personnel can be locked inside the store.

Freezing   34.15   Rapid and well-controlled freezing without delay after preparation is essential, at a rate appropriate to the specific product. Mechanical damage to the texture of cellular products can result from slow growth of large ice crystals.

34.16   Freezing equipment should be designed to give the minimum amount of product dehydration in order to prevent dehydration damage on the outer surface of the product.

34.17   Freezing should not be regarded as complete until a temperature of −18°C has been reached at the thermal centre of the food. The degree of homogeneity of the food needs to be considered especially if it is of a particulate nature of varying sizes. Freezer exit temperature should be such that the temperature of the products should not rise above −18°C during subsequent packing and palletisation. Some products can suffer surface cracking with too rapid freezing, for example, pastry products, and in this case, freezing will be a compromise of the process to give maximum safety control and quality to the centre phase while keeping the outer phase intact.

34.18   Food packaging should comply with the general requirements stated in Chapter 6, and also include a water vapour barrier to prevent dehydration and weight loss during storage.

34.19   Packaged frozen foods must carry the storage instructions necessary to validate the stated indication of minimum durability.

34.20   Precautions should be taken to ensure that packing integrity is maintained during processing, storage or distribution to avoid risk of product contamination, product rejection or freezer burn, which is caused by product dehydration.

34.21   When packing in bulk for later repacking, sacks or palletainers should be labelled with production date, process details, and variety and production batch to facilitate identifications at a later date. End-user packs should be labelled with repack dates (coded if necessary) and 'use before' dates. A record should be kept of goods that have been sent to individual repacking points so that end-user packs can be traced back to the actual production records if required (see Chapter 10).

Cold Chain   34.22   Primary long-term (3 months or more) stores should be maintained at or colder than −26°C. Temperature fluctuations of more than 2°C should be avoided and the frequency of variation kept to a minimum. The monitoring and recording of storage unit air temperature using calibrated temperature probes is essential. The number and location of temperature probes should be such as to ensure effective monitoring. Air temperatures in cold stores should be manually checked and logged at least once every 24 hours. Temperature warning and action limits should be specified and responsibilities for taking corrective action defined. Records should be maintained that detail any incident and the corrective action taken.

34.23   During delivery to the primary cold stores, products should not be exposed to direct sunlight, wind or rain; transfers should be carried out with the minimum of exposure to outside temperature conditions. Products incoming to the stores should not be warmer than −15°C and should remain in the stores until their temperature has reached −23°C or colder. Products leaving the stores should be protected as before from adverse weather or temperature conditions; as much as possible of the pre-removal operation should be carried out in the cold store. To afford maximum protection to the products during unloading and loading refrigerated transport, it is advisable to fit the stores with a loading bay with suitable protection positioned so that the temperature-controlled vehicle can be brought into direct contact with the store. The same principles should apply to retailers and caterers where there is a cold receipt area.

34.24   Secondary or distribution cold stores should be maintained at −23°C or colder. Temperature probes that monitor storage temperatures should be so positioned as to ensure effective monitoring. To this end, it is important that the relationships between core and surface temperatures are understood and their implications within the distribution and storage system taken into account. An alarm system should be operational to warn of any breakdowns. All autographic recorders and temperature probes should be accurate to within 1°C and ideally 0.5°C, and be calibrated at pre-defined intervals in such a way as to give traceability back to a national standard. If an autographic system is not available, manual logs should be maintained.

34.25   Cyclic variation of temperature, whether above or below the recommended temperature, is undesirable as it accelerates dehydration even in hermetically sealed packs when it causes migration of moisture from product to form 'snow' inside the packaging.

34.26   Clearly defined procedures should be laid down to deal with wide temperature fluctuations or refrigeration breakdown. Should the air temperature rise to −12°C, the cold store should be closed until the fault has been cleared and the correct store temperature restored. Products in the store during the period of breakdown should be checked by the quality control manager and records maintained of the decision for product disposition and associated actions including the removal of product to an alternative store.

34.27   Primary distribution vehicles should be designed so as to maintain the same temperature as the primary cold stores, and operated in such a way that the delivery temperature of the product is not higher than −18°C. If the products have a delivery temperature above −15°C, their quality should be checked, records maintained and appropriate action taken. Documented transport breakdown procedures should be in place, which define the actions to be taken in the event of vehicle or refrigeration unit breakdown, and the records to be maintained that detail the incident and the corrective action taken.

34.28   Secondary distribution vehicles are those used for transporting products between secondary cold stores and point of sale. They should be designed and operated in such a way that the food can be delivered at −15°C. If the product temperature is above −12°C, the overall quality should be checked, records maintained and appropriate action taken. Documented transport breakdown procedures should be in place, which define the actions to be taken in the event of vehicle or refrigeration unit breakdown, and the records to be maintained that detail the incident and the corrective action taken.

34.29   Temperature monitoring and recording in all distribution vehicles and holding centres should be such that relevant personnel are able to verify temperatures of vehicles and cold stores. Further recommendations are therein given for the transport, handling, storage and display by the wholesaler and retailer, and it is to the producer's ultimate advantage to encourage the observation of such recommendations.

General   34.30   Reference may also be made to the British Frozen Food Federation website for further guidance (http://www.bfff.co.uk) and to the **Ice**

Cream (Heat Treatment, etc.) Regulations 1959, SI 1959 No 734 as amended by SI 1962 No 1287 and SI 1963 No 1083, to the Quick-frozen Foodstuffs Regulations 1990, SI 1990 No 8136, as amended by SI 1994 No 298, implementing in the UK the EU Directive on the approximation of the laws of the Member States relating to quick-frozen foodstuffs for human consumption (89/108/EEC, 92/1EEC and 92/2/EEC).

35 DRY PRODUCTS AND MATERIALS

General

35.1 Processes involving dry materials have the problems associated with dust, particularly those of cleaning, the possible creation of an explosive dusty atmosphere and the risks of cross-contamination by dust particles. It is important therefore to contain dust as far as possible in an enclosed system and, with the aid of dust removal and extraction systems, to maintain a high standard of cleanliness. If allergens are utilised as raw materials, then dust control is critical, and control measures need to be in place according to the physical state of the material including particulate size and hardness of the particle.

35.2 The general environment of the plant and equipment including ledges and girders should be regularly cleaned, and an effective air extraction system should be installed. Such a system should discharge through a filter and at a point situated so as to minimise the risks of the discharge being able to contaminate the other plant or products. Dust extraction systems should be properly maintained, cleaned and serviced; they become heavily coated inside duct-work, and cleaning and filter changing can create a very dusty atmosphere. Dusty atmospheres should be considered as potentially dangerous explosion hazards. It may be desirable therefore to use flameproof motors and switches or ensure that they are situated in a relatively clean environment. Adequate protective clothing and other equipment should be provided for those involved physically in the cleaning operations, and during production if necessary.

35.3 Plant and equipment should be designed with easy access for cleaning in mind. When possible, manufacturing operations should be carried out in closed vessels or systems. Spray and fluid bed driers should be fitted with efficient filter bags. Where closed systems are not practicable, it is usually possible to carry out an operation within a dust extraction system. Delicacy of handling in relation to product friability may also need special attention.

Grinding, Mixing and Sieving

35.4 Particle size reduction of dry materials produces fine dust, and unless the process can be carried out in a closed system, including the discharge of the ground product, it should be operated within a dust extraction system. In flour mills where stone milling is undertaken, controls should be in place for the assessment of foreign body contamination risk. Further consideration should be given to the level of contamination present in wheat and the use of in-line magnets and screens to remove foreign bodies. Care must be taken to ensure that the product does not 'bridge' at these screening points. An investigation should be implemented where necessary to identify the likely source of the contaminant so that appropriate corrective action can be identified and instigated. Mixing and sieving operations usually

Food & Drink – Good Manufacturing Practice: A Guide to its Responsible Management, Sixth Edition.
The Institute of Food Science & Technology Trust Fund.
© 2013 John Wiley & Sons, Ltd. Published 2013 by John Wiley & Sons, Ltd.

213

have similar dust control problems, and these tasks should be undertaken using appropriate dust control methods.

General Hygiene	35.5	Dry goods production involves the same general hygiene considerations as other foods with regard to statutory requirements. These have been discussed in preceding chapters. A particular problem occurs with hygroscopic materials that become sticky and so call for special attention with regard to cleaning. They are not always completely removed by dust extraction systems, but those particles that are removed can clog ducts and filters. If a vacuum system is used for dry cleaning, then care must be taken to ensure that all pipework and attachments are stored in suitable locations in each production area and routinely inspected for signs of wear and damage and replaced as necessary. This inspection should be formally recorded. Wet cleaning methods should be available in all areas where such dry goods are produced and handled, but precautions should be taken against the risk of creating humid conditions, which might allow microbial growth.

Contamination	35.6	Dry materials require special attention concerning contamination and its detection. Particular attention should be given to examination for insect infestation, both past and present, and measures adopted to deal with either or both situations. At all times, dry materials should be protected against attack by insects and rodents. Dry goods should be inspected on arrival for signs of pest infestation including stored product insect contamination. There have been national recalls in the UK, for example, for both weevil and for biscuit beetle contamination, but other pests such as psocids (booklice), rust red flour beetle, flour moths and their larvae are also of concern. Visual inspection will include observing the materials, delivery vehicles and subsequently storage areas for signs of pests or infestation. Where materials/premises are found to be infested, the pest contractor must be informed and appropriate action taken. Any contaminated material must be disposed of immediately to prevent the further spread of contamination.

While the use of magnets has been previously discussed, all dry materials should pass through a metal detector at least once.

Microbiology	35.7	Most bacteria require a water activity (a_w) of at least 0.95, and very few microorganisms can grow in foods around 0.6. Some bacteria, for example, *Bacillus cereus*, can produce spores that will germinate on the reconstitution of the dry product. Yeasts and moulds can grow at lower levels of a_w.

	35.8	Dry foods and their ingredients can carry heavy microbiological loads undetected without the dry food undergoing any noticeable change. This can be exacerbated by incubative temperatures, which can occur during some drying and mixing operations. Vegetative organisms or spores can proliferate and cause problems when the dry food is reconstituted. Dried materials used as the ingredients in the manufacture of canned products should have low bacterial spore

counts. All ingredients, according to their origin and substance, should be subjected to a system of microbiological control by examination. Similarly, finished products should be sampled within a scheme of microbiological examination. When the food is to be consumed without further heat processing, it must be microbiologically safe. Particular care should be taken to prevent airborne dust from causing microbiological contamination of raw materials, finished product and plant. In areas that are dry-cleaned, microbiological contamination can be very difficult to remove.

Packaging and Storage

35.9 Consideration must be given to the potential minimum durability of dry products in the form in which they are offered for sale. Exposure of such a product to light, air and water vapour may cause physical or chemical or both types of change in the product, sometimes fairly rapidly, as may significant changes of temperature. Packaging can protect the product from these effects and can also protect from insects and rodents; it should therefore be considered and designed with these objectives in mind.

35.10 Where dry foods are intended to be reconstituted for consumption, detailed instructions for preparation given on the package should, where applicable, have regard to safety considerations.

36 COMPOSITIONALLY PRESERVED FOODS

36.1 Some foods, such as jams and pickled products, depend for their preservation and/or specific properties and maintenance of their quality during their expected life on achieving a particular quantitative composition [e.g. the attainment of requisite—and, in the UK and the other European Union (EU) Member States, also legally required—refractometric solids in jams with no added preservative, or of a preservation index of 3.6% acetic acid in the volatile constituents in non-pasteurised pickles and sauces].

36.2 In products where a quantitative compositional factor is critical, the training of production supervisors, operators and quality control staff should emphasise the critical nature of such compositional factors. Production methods and control procedures should be such as to ensure that the required composition is consistently achieved. Any production batch that is found to fall short should be quarantined, and dealt with in accordance with the procedures outlined in Chapter 21.

36.3 Where appropriate, relevant on-pack instructions for use should include instructions for storage.

Food & Drink – Good Manufacturing Practice: A Guide to its Responsible Management, Sixth Edition.
The Institute of Food Science & Technology Trust Fund.
© 2013 John Wiley & Sons, Ltd. Published 2013 by John Wiley & Sons, Ltd.

37   FOODS CRITICALLY DEPENDENT ON SPECIFIC INGREDIENTS

General

37.1   Some foods are critically dependent, for their preservation and/or specific properties and maintenance of quality during their expected life, on the presence in the required amount(s) of one or more particular specific ingredients, usually at relatively low levels. All product specifications should be validated as part of the design process before full production commences. Re-validation is required if there is a change to product formulation, ingredients supplier and so on to ensure that food safety will not be compromised (see Chapters 3 and 8). All ingredients must be declared 'food grade', and there must be documented evidence held by the manufacturer to substantiate this. Quantitative Ingredient Declarations (QUID) labelling must be complied with at all times for pre-packed foods (see 27.10).

37.2   Where the preservation and/or specific properties of the product critically depend on the inclusion, at a specified level, of one or more specific ingredients, the importance and critical nature thereof should be emphasised in the training of production supervisors, operatives and quality control staff. Production methods and control procedures should include specific provision directed towards ensuring that the substances concerned are not accidentally omitted, nor added in incorrect quantity, nor unevenly distributed through the product, whether resulting from operator error or inaccurate functioning of automatic dispensing equipment. Whereas a deficiency would impair the preservation and/or specific properties of the food, an excess might, in the case of certain specific additives, exceed specified legal limits and/or be harmful to the consumer.

37.3   Because of the wide variety of possible combinations of circumstances, it is impossible to generalise as to specific production techniques or control measures required for the purpose indicated in 37.2. Where discrete batch quantities of any specific ingredient are required, it may be appropriate for these to be accurately weighed by or under direct supervision of laboratory or quality control staff. Where colour or a coloured ingredient is also part of the product formulation, consideration should be given to the feasibility of pre-mixing or adding in the specific ingredient therewith, so that the visually obvious presence or absence of the colour in the batch acts as a 'marker' for the specific ingredients. Batch manufacturing records (see Chapter 9) should, of course, include provision for recording the addition of such substances, and operator-training should emphasise the importance of filling these in at the time of making the addition and not filling in a succession of ticks and initials at some later convenient time. It is very important that the operator

Food & Drink – Good Manufacturing Practice: A Guide to its Responsible Management, Sixth Edition.
The Institute of Food Science & Technology Trust Fund.
© 2013 John Wiley & Sons, Ltd. Published 2013 by John Wiley & Sons, Ltd.

initials the entry for each batch so that in the event of non-compliance, there is traceability to a particular operator (see 9.7). At the end of a production run, there should be a cross-check of the usage of the critical specific ingredients against the amount(s) required by the actual production volume during the run. Any discrepancy between the actual and expected amounts of a specific ingredient should be investigated.

Generally speaking, the natures of the substances likely to be involved do not lend themselves to continuous or rapid monitoring, so that analytical control provides only intermittent spot checks, which should, however, be carried out to confirm the effectiveness of the manufacturing disciplines on a retrospective basis. Chapter 10 deals with the importance of checking the identity of raw material supplies, that is, that they are in fact the substances that they purport to be. It is evident that this assumes additional importance in the case of the specific ingredients referred to above.

38   IRRADIATED FOODS

General

38.1   Food irradiation is the process of exposing food to a carefully controlled amount of ionising energy. The food is passed through a field of ionising energy from either machine-generated electron beams or gamma rays from cobalt-60. The ionising radiation passes through the food, generating large numbers of short-lived free radicals. These can kill microorganisms, such as *Salmonella*, and inhibit many processes, such as those that cause sprouting and ripening. At no time during the irradiation process does the food come into contact with the radiation source and, by using cobalt-60 or electron beams up to 10 MeV, it is not possible to induce radioactivity in the food. The length of time the food is exposed to the ionising energy and the strength of the source determine the irradiation dose, measured in kilograys (kGy), the food receives.

38.2   **The Food (Control of Irradiation) Regulations 1990 No 2490** introduce a strict system of licensing for food irradiation plants, with restrictions on food that can be irradiated and the dose permitted. The regulations also address the provisions for regular inspections and requirements for irradiated food, when stored or transported, to be accompanied by documentation identifying that it has been irradiated and providing for batch traceability. **The Food Labelling Regulations 1996** require that all irradiated foods or listed ingredients of foods are labelled with the words 'irradiated' or 'treated with ionising radiation'. **The Food Irradiation Provisions (England) Regulations 2000** implement two EU Directives on food irradiation. Similar regulations exist in Scotland, Wales and Northern Ireland.

Directive 1999/2/EC—Irradiation of Food and Food Ingredients lays down general provisions and the rules governing approval and control of irradiation and changes the rules on the labelling of foodstuffs that have been treated with ionising radiation. **Directive 1999/3/EC** establishes an initial positive list of foodstuffs that can be treated and freely traded within the EC. **Document 2001/C 128/08** has been implemented, identifying the list of Member States authorisations of food and food ingredients that may be treated with ionising radiation. **Document 2001/C 128/09** provides for the list of approved facilities for the treatment of food and food ingredients with ionising radiation in the Member States. **The Food Irradiation (England) Regulations 2009**[1] or its equivalent implementing **EU Directive 1999/2/EC** and **1999/3/EC**, as amended by the **Commission Decision 2002/840/EC, 2004/691/EC** and **2007/802/EC** and finally **2010/172/EU**, contains mandatory labelling requirements on irradiated food sold. The **Food Irradiation (England) (Amendment) Regulations 2010** made further amendments to the legislation.

[1]http://www.legislation.gov.uk/uksi/2009/1584/pdfs/uksi_20091584_en.pdf.

Food & Drink – Good Manufacturing Practice: A Guide to its Responsible Management, Sixth Edition.
The Institute of Food Science & Technology Trust Fund.
© 2013 John Wiley & Sons, Ltd. Published 2013 by John Wiley & Sons, Ltd.

38.3   The following types of ionising radiation may be used:

(a) gamma rays from the radionuclides cobalt-60 or caesium-137;
(b) X-rays generated from machine sources at or below an energy level of 5 million electron-volts (MeV);
(c) electrons generated from machine sources at or below an energy level of 10 MeV.

38.4   Exemption from the UK regulations is given to those foods exposed to low-radiation doses [less than 0.5 gray (Gy)] from measuring or detection instruments used for inspection purposes. Foods that are radiation sterilised for hospital patients' diets are also exempt.

38.5   In the UK, only foods that fall within the following classes can be considered for irradiation:

Maximum overall average dose in kilograys (kGy)[2]

Fruit (including fungi, tomatoes, rhubarb)	2
Vegetables (including pulses)	1
Cereals	1
Bulbs and tubers (potatoes, yams, onions, shallots, garlic)	0.2
Dried aromatic herbs, spices and vegetable seasonings	10
Fish and shellfish	3
Poultry	7

These categories of food can first be irradiated and then used as ingredients in other food products. A mixture of foods from the same class, for example, a blend of herbs and spices, can be irradiated. A composite food product is not permitted for irradiation treatment unless minor ingredients (including additives) together comprise no more than 2% by weight, excluding any added water from the calculation.

Supplier approval of ingredients should pay particular attention to whether spices have been irradiated at source not only to ensure appropriate labelling of the finished product but also as part of raw material risk assessment. *Salmonella* contamination of foods has been linked to ingredients such as ground black pepper.

38.6   The approved purposes of irradiation of food are:

(a) elimination or reduction of pathogenic organisms in food;
(b) reduction of spoilage of food by retardation or arresting of decay processes and destruction of spoilage organisms;
(c) reduction of waste food resulting from premature ripening, germination or sprouting; and
(d) disinfection of food from infestation by organisms harmful to plants or plant products.

---

[2]http://www.food.gov.uk/safereating/irradfoodqa/.

38.7 The overall average dose calculation is defined in the **Food Irradiation (England) Regulations 2009**; however it should be noted that while the dose can be readily determined for homogenous products, it is more difficult in a heterogeneous product where particulate size or density varies. In some cases, the mean value of the average values of the minimum dose measured in the product (D_{min}) and the maximum dose (D_{max}) will be a good estimate of the overall average dose. In this instance, the overall average dose = $(D_{max} + D_{min})/2$. The ratio of D_{max}/D_{min} (dose uniformity ratio) must not exceed 3 and the maximum absorbed dose must not exceed 150% of the overall average dose.

Facilities and Control of the Process

38.8 Irradiation treatment of foods must be carried out in facilities licensed and registered for this purpose by the appropriate authority. Irradiation procedures must be fully validated and records retained to demonstrate their efficacy and their compliance with legislation. Re-validation is required whenever the product, its geometry or the irradiation conditions are changed. This re-validation process must be formally recorded.

During the process, routine dose measurements are carried out in order to ensure that the dose limits are not exceeded. Measurements should be carried out by placing dosimeters at the positions of the maximum or minimum dose, or at a reference position. The dose at the reference position must be quantitatively linked to the maximum and minimum dose. The reference position should be located at a convenient point in or on the product, where dose variations are low.

Routine dose measurements must be carried out on each batch and at regular intervals during production. In cases where flowing, non-packaged goods are irradiated, the locations of the minimum and maximum doses cannot be determined. In such a case, it is preferable to use random dosimeter sampling to ascertain the values of these dose extremes. Dose measurements should be carried out by using recognised dosimetry systems, and the measurements should be traceable to primary standards.

During irradiation, certain facility parameters must be controlled and continuously recorded. For radionuclide facilities, the parameters include product transport speed or time spent in the radiation zone and positive indication for the correct position of the source. For accelerator facilities, the parameters include product transport speed and energy level, electron current and scanner width of the facility.

38.9 The irradiation facility must be licensed for specific foods at specific doses (see 38.5).

38.10 A person seeking a licence to irradiate food should apply, giving the following details with the written application:

(a) the applicant's name;
(b) the applicant's address;

(c) the address of the facility at which the applicant proposes to irradiate food;

(d) details of any licence or registration under any other legislation that enables the applicant to use ionising radiation at the facility in circumstances where, but for that licence or registration, that use would be unlawful;

(e) a description of each food that the applicant proposes to irradiate which is sufficient to show that it falls within a permitted category of food;

(f) in respect of each food described pursuant to sub-paragraph (e) –

    (i) the purpose for which the applicant proposes to irradiate the food and how that would benefit consumers;

    (ii) the method by which the applicant will ensure that the food is in a suitably wholesome state before irradiation;

    (iii) the overall average dose, maximum dose and minimum dose of ionising radiation that the applicant proposes to apply to the food;

    (iv) the method (including instrumentation and frequency) by which the applicant proposes to measure any dose of ionising radiation and the dosimetry standard that the applicant proposes to use to calibrate the dose meters used to measure it;

    (v) whether or not the applicant proposes to irradiate that description of food in packaging in contact with the food and, if so, the packaging that the applicant proposes to use; and

    (vi) whether or not the applicant proposes to apply temperature control to the food while irradiating it and, if so, the temperature at which the applicant proposes to keep the food during the application of temperature control.

38.11 The facilities must be designed to meet the requirements of safety, efficacy and good hygienic practices of food processing, in accordance with current statutory regulations in the UK and elsewhere and good manufacturing practice (GMP). A list of licenced facilities is contained in Schedule 1 and Schedule 2 of the **Food Irradiation (England) (Amendment) Regulations 2010**.[3]

38.12 Control of the process involves the food manufacturer and the operator of the irradiation facility (licence holder). Controls within the facility must include keeping records such as licence details, treatment and type of each food consignment, plus details of dose measurements. Additionally, each consignment must be accompanied throughout the processing chain by relevant documentation. Premises and records are open to inspection by the licensing authority. The licence holder is responsible for retaining records for a minimum of 5 years.

38.13 Incoming product must be kept physically separate from the outgoing product.

[3]http://www.legislation.gov.uk/uksi/2010/2312/pdfs/uksi_20102312_en.pdf.

Food Quality   38.14   The food before and after irradiation should comply with the provisions of the Codex Alimentarius Recommended International Code of Practice—General Principles of Food Hygiene and, where appropriate, with the Codex Recommended International Code of Hygienic Practice relative to the particular food.

38.15   Food intended for irradiation should be of a similar microbiological quality to that being processed by other means. For those foods irradiated to eliminate or reduce pathogenic organisms in the food, details of microbiological testing are required. **EU Regulation No 2073/2005 on microbiological criteria for foods (as amended by EU Regulation No 1441/2007)**[4,5] complements EU food hygiene legislation already described and applies to all food businesses involved in the production and handling of food. The microbiological criteria outlined can be used to validate manufacturing, handling and distribution processes and/or verify the acceptability of foodstuffs at specific stages in the process. The Food Standards Agency (FSA) has produced general guidance for food business operators.[6]

38.16   In the UK, public health requirements affecting microbiological safety and nutritional adequacy of specific foods must be observed.

38.17   Irradiation must not be used to process foodstuffs that do not meet the above microbiological quality standards.

38.18   Documentation referring to microbiological quality should accompany the food consignment from the manufacturer to the irradiation plant.

Application of the   38.19   Irradiation should only be applied to food manufactured in accordance with GMP, and with a demonstrable need for the particular use of the process in terms of food hygiene or other technological benefit (see 38.6). The doses applied should be commensurate with the application and within the set limit for overall average dose for the food (see 38.5). Food manufacturers should define the correct dose to be applied, which should be adhered to by the irradiation plant operator.

38.20   Temperature control during distribution, handling and irradiation processing should be similar to the control exerted on analogous but non-irradiated foods. Irradiated food must be stored and transported with appropriate documentation that contains the statement to the effect that the food has been treated with ionising radiation.

38.21   Packaging materials should be of suitable quality and acceptable hygienic condition and be appropriate for the purpose. They should be able to withstand radiation in terms of their physical and sensory

[4]http://eur-lex.europa.eu/LexUriServ/LexUriServ.do?uri=CONSLEG:2005R2073:20071227:EN:PDF.
[5]http://eur-lex.europa.eu/LexUriServ/LexUriServ.do?uri=OJ:L:2007:322:0012:0029:EN:PDF.
[6]http://www.food.gov.uk/multimedia/pdfs/ecregguidmicrobiolcriteria.pdf.

properties and in terms of migration of mobile components into the food. A full description of the materials is required. Packaging dimensions, particularly those of bulk packs, should take into consideration the nature of the irradiation source with respect to depth of penetration.

Re-Irradiation	38.22	Food must not be re-irradiated. However, the full dose needed for a specific technological purpose may be given in sequential fractionated doses.
Import	38.23	Imports of irradiated foods into the UK are permitted only from countries that have been approved by the licensing authority. Lists of irradiation plants approved to supply specific irradiated foods and lists of countries approved to export specific irradiated food should be consulted. Appropriate documentation must accompany each consignment.
Export	38.24	Foods irradiated for export should comply with the guidelines described in Chapter 43.
	38.25	Where the importing country's statutory requirements differ from the exporting country's standards, it is the responsibility of the suppliers to determine that their irradiated raw material or product will comply with these.
Labelling	38.26	In the UK, irradiated foods, whether pre-packaged or not, must be labelled with the statement 'irradiated' or 'treated with ionising radiation'. This labelling requirement includes foods sold in catering establishments. Any irradiated ingredient incorporated into a food product necessitates the use of the same labelling terms in the ingredients list (see relevant food labelling regulations).

39   NOVEL FOODS AND PROCESSES

General

39.1   Care should be taken in the use of novel food or food ingredients produced from raw material that has not hitherto been used (or has been used only to a small extent) for human consumption in the area of the world in question, or that is produced by a new or extensively modified process not previously used in the production of food (and would thus include genetically modified materials). This must include attention to food safety considerations, compliance with the relevant regulations of the country for which the food is intended, and provision of label information to enable the purchaser or consumer to make an informed choice.

39.2   Novel foods and processes, including especially genetic modification (GM), have the potential to offer very significant improvements in the quantity, quality and acceptability of the world's food supply. Food scientists and technologists can support the responsible introduction of such foods and processes provided that issues of product safety, environmental concerns, information and ethics are satisfactorily addressed.

39.3   In the European Union (EU), the principal legislation is **EU Regulation No 258/97** of the European Parliament and of the Council of 27 January 1997 concerning novel foods and novel food ingredients (OJ L43, 14.2.97, pp. 1–7). Enforcement and execution of certain provisions of the EU Regulation and set fees for assessment procedures in the UK are provided by **The Novel Foods and Novel Food Ingredients Regulations 1997 (SI 1997/1335) and The Novel Foods and Novel Food Ingredients (Fees) Regulations 1997 (SI 1997/1336)**. However, any release of genetically modified organisms (GMOs) into the environment is now governed by national regulations implementing **EU Directive 2001/18/EC** [in the UK by the **Genetically Modified Organisms (Deliberate Release) Regulations 2002**], subject to amendments to the directive consequent on **Regulation (EC) 1829/2003 on Genetically Modified (GM) Food and Feed**; **Regulation (EC) 1981/2006** and **Regulation (EC) 298/2008**.

The **Novel Foods and Novel Food Ingredients Regulations (1997)** applied to the placing on the market within the EU of foods and food ingredients that have not previously been used for human consumption to a significant degree and that fall into the following categories:

- foods and food ingredients containing or consisting of GMOs within the meaning of **Directive 90/220/EEC** (see below), now superseded as above;
- foods and food ingredients produced from, but not containing, GMOs, foods and food ingredients with a new or intentionally modified primary molecular structure, now superseded as above;

Food & Drink – Good Manufacturing Practice: A Guide to its Responsible Management, Sixth Edition.
The Institute of Food Science & Technology Trust Fund.
© 2013 John Wiley & Sons, Ltd. Published 2013 by John Wiley & Sons, Ltd.

- foods and food ingredients consisting of or isolated from micro-organisms, fungi or algae;
- foods and food ingredients consisting of or isolated from plants and food ingredients isolated from animals, except for foods and food ingredients obtained by traditional propagating or breeding practices and having a history of safe food use;
- foods and food ingredients to which has been applied a production process not currently used, where that process gives rise to significant changes in the composition or structure of the foods or food ingredients that affect their nutritional value, metabolism or level of undesirable substances.

The novel foods and food ingredients must not present a danger for the consumer, mislead the consumer or differ from the foods or food ingredients that they are intended to replace to such an extent that their normal consumption would be nutritionally disadvantageous for the consumer. Derogations are available for foods and food ingredients that, according to expert scientific opinion, are substantially equivalent to existing foods in respect of their composition, nutritional value, metabolism, intended use and level of undesirable substances contained therein.

39.4   The **EU Novel Foods Regulation** also specifies the assessment procedures that must be carried out before a novel food can be placed on the market, and makes provision for objections to be raised by interested parties. A procedure for re-assessment of novel foods is included if they subsequently appear to be endangering human health and for a review of the regulations within 5 years from implementation in any case.

39.5   The European Food Safety Authority (EFSA) is now the central body controlling the assessment of novel foods and novel food ingredients. Applications can be made to a national authority [in the UK, the Advisory Committee on Novel Foods and Processes (ACNFP)]. Any person or company contemplating marketing in the UK a novel food or one containing a novel ingredient that has not already been the subject of official evaluation and approval **must make a prior submission to the ACNFP.**

39.6   **Council Directive 90/220/EEC** of 23 April 1990 on the deliberate release into the environment of GMOs [OJ L117, 8 May 1990, p. 15; amended by the **Commission Directive 94/15/EC** (OJ L103, 2214/94, p. 20) and **Commission Directive 97/35/EC** (OJ L169, 27/6/97, p. 72)]. This directive is implemented in the UK by the following regulations: **The Genetically Modified Organisms (Deliberate Release) Regulations 1992 (SI 1992/3280) and The Genetically Modified Organisms (Deliberate Release) Regulations 1995 (SI 1995/304) and The Genetically Modified Organisms (Deliberate Release and Risk Assessment—Amendment) Regulations 1997 (SI 1997/1900).** Under this directive, a manufacturer or importer must submit a notification to the national competent body of a Member State where the product is to be first placed on the market, before undertaking a deliberate release into the environment of a GMO, or placing it on the market.

The notification should contain a technical dossier of information including a full risk assessment. The Member State that received the notification examines the dossier and in the case of a negative evaluation, the notification is rejected. In the case of a favourable opinion, the dossier is forwarded to the European Commission and all the competent authorities of the other Member States who have the right to raise objections. If there are no objections, the competent authority that carried out the original evaluation grants the consent for the placing on the market of the product, which may then be placed on the market throughout the EU. In case of objections, a decision has to be taken at community level. The commission seeks the opinion of the EFSA before drafting a decision, which is put forward to the regulatory committee composed of representatives of Member States for favourable opinion. Otherwise a proposal is put forward to the council, which decides by qualified majority. If no council decision is taken within 3 months, the commission takes the decision. In any case, in accordance with Directive **90/220/EEC**, the commission is ultimately obliged to adopt measures to authorise a GMO, if the application fulfils current EU legislation and if it is not rejected by unanimity in the council, or if the council fails to act within the fixed deadline.

39.7   GMOs have to undergo a scientific assessment of risks to human health and the environment before receiving community authorisation. Risk assessments are performed on a case-by-case basis. The safety assessment takes account of the following:

- how the GMO was developed – including the source of the genes to be introduced and detailed molecular analysis of the modified plant and organism. The process can be likened to that of 'cutting' and 'pasting' where pieces of DNA are cut out of the donor organism, copied and pasted into a recipient organism. It is necessary to establish which genes are incorporated and where in the recipient genome.
- risk associated with the gene products in the plant, mainly proteins. It is necessary to know that the gene does not encode for a protein that is toxic to humans or does not produce an allergic response. It must also be established that the inserted gene(s) does not result in unexpected effects.
- investigation of the possibility that the inserted gene may be transferred to bacteria. This has particular relevance to the possible transfer of antibiotic resistance genes.

39.8   **EU Council Directive 90/219/EEC on the contained use of genetically modified microorganisms (OJ L117, 8 May 1990, pp. 1–14), as amended by Council Directive 98/81/EC of 26/10/98 (OJ L330, 5 December 1998, p. 13)**. This directive, which provides the circumstances and conditions under which GMOs (including fermentation organisms) require consent for contained use, is implemented in the UK by:

- **The Environmental Protection Act 1990, Part VI, Genetically Modified Organisms, Sections 106–127**. Section 106 states that this part (i.e. Part VI) has effect for preventing or minimising any damage to the environment, which may arise from the escape or release from human control of GMOs;
- **The Genetically Modified Organisms (Risk Assessment) (Records and Exemptions) Regulations 1996 (SI 1996/1106) restrict the import and acquisition of GMOs under Section 108 (l) (a) of this act**;
- **The Genetically Modified Organisms (Contained Use) Regulations 1992 (SI 1992/32 17) [The Genetically Modified Organisms (Contained Use) Regulations 1993—now revoked]. The Genetically Modified Organisms (Contained Use) (Amendment) Regulations 1996 (SI 1996/967) and The Genetically Modified Organisms (Contained Use) (Amendment) Regulations 1996 (SI 1998/1548)**.

39.9   Regardless of the scientific validity or otherwise of the reasons for which individual purchasers make their choices, they should be provided with the means of making an informed choice. This involves a combination of (a) information about what the food contains or the novel process by which it was made, and (b) an understanding of the significance of that information. (a) must be provided on the label and (b) should be summarised there if the purchaser would be disadvantaged by failing to do so.

39.10   Labelling legislation was approved by the Council of Ministers on 22 July 2003, published in the Official Journal of the European Union on 22 September 2003 and designated **Regulation (EC) 1829/2003** of the European Parliament and of the Council of 22 September 2003 on genetically modified food and feed and **Regulation (EC) 1830/2003** of the European Parliament and of the Council of 22 September 2003 concerning the traceability and labelling of GMOs and the traceability of food and feed products produced from GMOs and amending Directive 2001/18/EC.

39.11   In the UK, the Food Standards Agency (FSA) has carried out a consultation on the enforcement and penalties in the UK of these two EU regulations, and on draft guidance notes,[1] although, as FSA points out, this is a draft informal guidance for the application of both regulations in the UK and cannot modify the regulations themselves. In the UK, the Department of Environment, Food and Rural Affairs (DEFRA) carried out a consultation on proposals for administering and enforcing Regulation (EC) No 1946/2003 of the European Parliament and of the Council on transboundary movements of GMOs, which came into force in November 2004.

39.12   For the latest Institute of Food Science & Technology (IFST) information statement, follow the link below.[2]

[1]http://www.food.gov.uk/multimedia/pdfs/gmconsultnotes.pdf.
[2]http://www.ifst.org/science_technology_resources/for_food_professionals/information_statements/.

40   FOODS FOR CATERING AND VENDING OPERATIONS

General   40.1   The manufacture of foods intended for use in catering or vending operations should be carried out in accordance with the principles and practices outlined in this Guide but, additionally, should have regard to the special requirements that relate to the intended use. It should be noted that 'manufacture of food' in this context applies not only to food products made by a food manufacturing company and sold to a caterer or to a vending machine operator, but also to food prepared in a central production unit by factory-style processing, by a catering organisation for use in its own catering establishments, as distinct from preparation in 'cook-serve' form.

40.2   In the manufacture and distribution of food and drink for catering and vending purposes, particular regard should be given to the circumstances and conditions of use, the probable expertise of the caterer and his/her staff, and the interactions likely to occur between the product and its subsequent environment. The general points of guidance identified as pertinent to good manufacturing practice (GMP) apply equally to foods manufactured for catering and vending purposes. The manufacturer should be prepared to offer technical advice to users on the suitability of products for the uses intended and on any appropriate precautions to be observed.

40.3   Adequacy of information (such as ingredients and nutrition information) and its intelligibility to the intended user are of particular importance. This includes the need to recognise possible literacy and language problems, stock rotation and effective product life, and appropriateness of presentation and packaging (e.g. in-pack microwaveability, or product stability during prolonged maintenance at serving temperature and relative humidity).

40.4   Where a catering organisation is preparing food by factory-style cook-freeze or cook-chill processing in the UK, reference should also be made to the Food Standards Agency (FSA) guidelines or further information is to be found at http://www.campden.co.uk/publications/pubs.php.

Vending   40.5   The interaction effects between product, environment and equipment are especially pertinent in operations involving vending machines. In the manufacture of products for these purposes, the manufacturer should ensure awareness of such potential hazards as within-machine environment, hygiene and cleaning needs, product flow properties, variability of throw or dispensation, as well as product and machine interactions and interactions with other products or ingredients.

Food & Drink – Good Manufacturing Practice: A Guide to its Responsible Management, Sixth Edition.
The Institute of Food Science & Technology Trust Fund.
© 2013 John Wiley & Sons, Ltd. Published 2013 by John Wiley & Sons, Ltd.

40.6   The requirements for vending operations may call for particular product performance standards, for example, dispersion at sub-scalding temperatures or interchangeability with competitors' products in an identical vending situation.

40.7   Reference may also be made to the Automatic Vending Association (AVA) publication Guide to Good Hygiene Practice in the Vending Industry.[1] The AVA has also developed their own AVA Quality System that their members must comply with. Further details can be found at http://www.ava-vending.co.uk and the European Vending Association (EVA) at http://www.eva.be.

Complaints   40.8   Manufacturers should develop appropriate systems to handle record and respond to complaints from caterers, and from their customers via caterers, as these will not normally fit into the system for dealing with direct complaints from consumers who have purchased retail products.

[1]http://79.170.40.235/ava-vending.co.uk//media/hygiene-guidelines(1).pdf.

41   THE USE OF FOOD ADDITIVES AND PROCESSING AIDS

41.1   The manufacturer should satisfy himself/herself that the use of a processing aid serves an essential functional need in manufacture, and that the inclusion of any additive serves an essential technical need or contributes significantly to customer quality requirements, or both.

41.2   The **Food Additives Labelling Regulations 1992** implement provisions in **Council Directive 89/107/EEC** on the approximation of the laws of the Member States concerning food additives authorised for use in food intended for human consumption has been repealed. **EC Regulations 1333/2008 and 1129/2011** are now in force. The regulations are amended by the **Sweeteners in Food Regulations 1995 No 3123, the Colours in Food Regulations 1995 No 3124, the Miscellaneous Food Additives Regulations 1995 No 3187, the Food Labelling Regulations 1996 No 1499, the Miscellaneous Food Additives (Amendment) Regulations 1999 No 1136 and the Miscellaneous Food Additives (Amendment) (England) (No 2) regulations 2001 No 3775.**

The use of any food additive or processing aid substance must comply with the laws and specific regulations of the country for which the food is intended. **The UK Colours in Food Regulations 1995, SI 1995 No 3124 implement in the UK the EU Parliament and Council Directive 94/36/EC and the Commission Directive 95/45/EC. The Colours in Food (Amendment) (England) Regulations 2000 No 481** amend these to cover **Commission Directive 1999/75/EC**, which changed the specification for E160a (i). **The Colours in Food (Amendment) (England) Regulations 2001 No 3442 (and similar regulations for Scotland, Wales and Northern Ireland)** further amend the **1995 Regulations** in light of **Commission Directive 2001/50/EC.**

The UK Flavourings in Food Regulations 1992, SI 1992 No 1971, as amended by the UK Flavourings in Food (Amendment) Regulations 1994, SI 1994 No 1486, implement in the UK the EU Council Directive 88/388/EEC as amended by Commission Directive 91/71/EEC.

The UK Miscellaneous Food Additives Regulations 1995, SI 1995 No 3187, as amended by the UK Miscellaneous Additives (Amendment) Regulations 1997, SI 1997 No 1413, implement in the UK the EU Parliament and Council Directive 95/2/EC and EC Directives 95/2/EC 96/77/EC and 96/85/EC. The Miscellaneous Additives (Amendment) Regulations 1999, No 1136 implement

Food & Drink – Good Manufacturing Practice: A Guide to its Responsible Management, Sixth Edition.
The Institute of Food Science & Technology Trust Fund.
© 2013 John Wiley & Sons, Ltd. Published 2013 by John Wiley & Sons, Ltd.

European Parliament and Council Directive 98/72/EC and Commission Directive 98/86/EC. The Miscellaneous Additives (Amendment) (England) Regulations 2001 No 60 (and similar regulations for Scotland, Wales and Northern Ireland) and implement Commission Directive 2000/63/EC. The Miscellaneous Additives (Amendment) (England) (No 2) Regulations 2001 No 3775 (and similar regulations for Scotland, Wales and Northern Ireland) and implement Commission Directive 2001/30/EC and 2001/5/EC. The Miscellaneous Additives (Amendment) (England) Regulations 2003 No 1008 implement Commission Directive 2002/82/EC.

The UK Sweeteners in Food Regulations 1995, SI 1995 No 3123, as amended by the UK Sweeteners in Food (Amendment) Regulations 1996, SI 1996 No 1477 and SI 1997 No 814, implement in the UK the European Parliament and Council Directive 94/35/EC and Commission Directive 95/31/EC, as amended by EC Directive 96/83/EC. The UK Sweeteners in Food (Amendment) Regulations 1997 No 314 implement EC Directive 96/83/EC. The UK Sweeteners in Food (Amendment) Regulations 1999 No 982 implement EC Directive 98/66/EC. The UK Sweeteners in Food (Amendment) (England) Regulations 2001 No 2294 implement EC Directive 2000/51/EC with regard to E421 and E965 (ii). The UK Sweeteners in Food (Amendment) (England) Regulations 2002 No 379 implement EC Directive 2000/52/EC with regard to E421 and E951 with further amendments in the UK Sweeteners in Food (Amendment) (England) Regulations 2003 No 1182.

41.3 Guidelines both for the principles that governments should adopt in approving the use of specific additives and for the considerations with which manufacturers should comply in using additives, were established in 1956 by the Joint FAO/WHO Expert Committee on Food Additives (JECFA) and have subsequently been embodied in **Section 5 of the Codex Alimentarius General Principles and in Annex 11 of EU Directive 89/107/EEC**. The principles involved are summarised in the following Sections 41.4–41.7.

41.4 Governments and their expert advisers must consider both need and toxicological evaluation to establish safety-in-use before approving an additive, and must review such approval on a continuing basis. These principles already apply in the UK and most other countries.

41.5 In the UK and other European Union (EU) Member States, where a maximum level of usage of any substance is specified in the appropriate directives and regulations listed in Section 41.2, the level must not be exceeded; but, additionally, the level of usage of a specific substance should not exceed the lowest level that produces the required effect when used efficiently in good manufacturing practice.

41.6 The use of an additive must not mislead consumers as to the quality of a product. This necessitates appropriate labelling that includes information about additives present, given in accordance with the regulations of the country for which the food is intended. It also implies that the labelling and advertising language or pictorial representation should not be used to create a false or misleading impression.

41.7   Additives should not be used to permit or to disguise the effect of faulty processing.

42   RESPONSIBILITIES OF IMPORTERS

Note: For the purposes of this Guide, 'import' may include the movement of goods between Member States of the European Union or within similar trading regimes.

42.1   There is a responsibility on the importer of foods and drinks to satisfy himself/herself and the appropriate authorities that the imported products have been produced in accordance with the principles of good manufacturing practice (GMP) and that they, and their mode of packaging and distribution, comply with the relevant legislative requirements at the points of import and sale.

42.2   In all respects, the importer should ensure that the requirements outlined in this guide are met as though (s)/he were, herself/himself, the producer of the products (for the import of food into the UK, attention is drawn to the due diligence aspects of the **Food Safety Act 1990**).

42.3   The importer should ensure that all imports are obtained against a clear and legally valid product specification. Attention is drawn to a UK legal requirement to obtain appropriate certification in certain instances. In this connection, from 1 January 1998, UK importers of meat, meat preparations, meat products or products containing gelatine or tallow, need with each consignment a certificate in accordance with the **Specified Risk Material Order 1997**, and signed by the veterinary authority in the country of despatch.

42.4   Wherever possible, inspection should be carried out at the point of origin to ensure that the agreed specification is met.

42.5   The goods on receipt in the country of destination should be subjected to a quality control evaluation that should take due account of any changes or damage that might have occurred in transit, and that should reflect the confidence level estiablished in the competence of the supplier and the current and historic availability of data on product quality and integrity from that origin. On importation, all the relevant legislative requirements, including those pertaining to labelling and metrology, must be met.

42.6   Consideration should be given to the requirements of Trading Standards and Port Health Authorities and the possibility of problems arising from their needs to examine and clear shipments. Close liaison with officers is to be commended, particularly with regard to

Food & Drink – Good Manufacturing Practice: A Guide to its Responsible Management, Sixth Edition.
The Institute of Food Science & Technology Trust Fund.
© 2013 John Wiley & Sons, Ltd. Published 2013 by John Wiley & Sons, Ltd.

sampling, clearance of perishable goods, identification requirements and any special needs relating to the product in question.

42.7 The importer should ensure adequate liaison with the manufacturer on matters relating to legislative or other changes in requirements for the products, and exchange information resulting from complaints, the needs of his/her own customers and the ultimate consumers.

43   EXPORT

Note: For the purposes of this Guide, 'export' may include the movement of goods between Member States of the European Union or within similar trading regimes.

43.1   The manufacture of foods for export purposes requires the general compliance with the principles embodied throughout this Guide, namely those of meeting specific legislation and/or regulatory obligations, and of ensuring that packaging and shipping arrangements are adequate and suitable.

43.2   It is the responsibility of the producer, unless manufacturing for a specific contract on behalf of the importer or ultimate customer, to be aware of and follow the pertinent rules relating to export from the country of origin and to acceptance in the country of destination. For example, licensing or inspection of the producing plant may be required, as may certification of materials or processes. In some cases, inspection or enforcement may lie with the authorities from the country of destination, or their agents, while in others, national or local authorities may carry out an acceptance function.

43.3   Due regard should be paid to the requirements of Port Health Authorities and Customs. See also Chapter 42 (42.6).

43.4   Attention should be paid to the packaging and shipping requirements, particularly with regard to dimensions, weights, stowage, protection, hygiene and contamination risks. Where appropriate, clear instructions should be provided on the packaging and/or in accompanying documentation, on any special treatment or precautions required and procedures to adopt in the event of breakage, accident or delay.

43.5   Due attention to use of appropriate languages and international symbols should be given in relation to transit arrangements and ultimate destination.

43.6   In the manufacture of products for export, due regard should be paid to the suitability of the product formulation and packaging for the environmental circumstances likely to be encountered en route and at the ultimate destination.

Food & Drink – Good Manufacturing Practice: A Guide to its Responsible Management, Sixth Edition.
The Institute of Food Science & Technology Trust Fund.
© 2013 John Wiley & Sons, Ltd. Published 2013 by John Wiley & Sons, Ltd.

PART III – MECHANISMS FOR REVIEW OF THIS GUIDE

This Guide is updated from time to time to incorporate any changes found to be necessary in the light of experience, changing legislation or advances in food science and technology and to correct errors or omissions and so on.

Food & Drink – Good Manufacturing Practice: A Guide to its Responsible Management, Sixth Edition.
The Institute of Food Science & Technology Trust Fund.
© 2013 John Wiley & Sons, Ltd. Published 2013 by John Wiley & Sons, Ltd.

APPENDIX I

DEFINITION OF SOME TERMS USED IN THIS GUIDE

Analytical Method A detailed description of the procedures to be followed in performing tests for assessing conformity with the specification.

Aseptic Processing A production process where a commercially sterile product (see Commercial sterility) otherwise known as aseptic product is packed into a container that has been independently and previously sterilised in a way that prevents contamination and maintains full sterility. Examples include Tetra Pak technology and Bag-in-Box technology.

Batch The quantity of material, which has been produced during a defined period of manufacture. A 'batch' may actually have been produced by a batch-wise process, or may correspond to a particular time duration during the run of a continuous process.

Batch Manufacturing Record A document stating the materials used and operations carried out during the manufacture of a given batch, including details of in-process controls and the results of any corrective action taken. It should be based on the master manufacturing instructions, and be compiled as the manufacturing operation proceeds.

Batch Number A unique combination of numbers or letters, or both, used to identify a batch and permit its history to be traced.

Botulinum Cook The heat treatment given to a low-acid canned food (having a pH higher than 4.2) sufficient to inactivate 10^{12} spores of *Clostridium botulinum*. This heat treatment is called the F_o value, and it is equivalent to a process of 3 minutes at 121°C, 10 minutes at 115°C or 32 minutes at 110°C.

Bulk Product Any product that has completed all processing stages up to, but not including, packaging (not applicable to those products where processing takes place inside the container and the latter is itself therefore part of the processing).

Chill Chain An organised system governing the conditions under which chilled foods are stored and handled by the producer, distributor and retailer. The conditions, as set out in Chapter 33 (33.1, 33.22–33.23), are those which ensure that temperatures maintained during storage, distribution and sale are those consistent with maintenance of quality and safety.

Chilled Foods Perishable foods that, to extend the time during which they remain wholesome, are kept within controlled and specified ranges of temperature above their freezing points and normally below 8°C.

Food & Drink – Good Manufacturing Practice: A Guide to its Responsible Management, Sixth Edition.
The Institute of Food Science & Technology Trust Fund.
© 2013 John Wiley & Sons, Ltd. Published 2013 by John Wiley & Sons, Ltd.

Cold Chain	An organised system governing the conditions under which frozen foods are stored and handled by the producer, distributor and retailer. The conditions, as set out in Chapter 34 (34.22–34.29), are those which ensure that temperatures maintained during storage, distribution and sale are those consistent with maintenance of quality and safety.
Commercial Sterility	A term common in the canning industry meaning the condition achieved by the application of heat that renders the processed product free from viable microorganisms, including those of known public health significance, capable of growing in the food at the temperatures at which the food is likely to be held during distribution and storage.
Contract Manufacture	Manufacture or partial manufacture ordered by one person or organisation (the contract giver) and carried out by a separate person or organisation (the contract acceptor).
Corrective	The action taken on the identification of non-conformance in terms of ingredients, products, behaviours or activities with specified documented requirements.
Critical Control Point	CCP; a material, or a location, or a practice, or a procedure, or a process stage where loss of control would result in an unacceptable food safety risk.
Dehydrated Food	Food or food products from which all but a small percentage of the water has been removed under controlled conditions.
Detergent	A chemical or mixture of chemicals that are used to remove soiling or grease from a surface, leaving it accessible to the action of disinfectants.
Disinfection	Defined by British Standard 5283:1986 as 'the destruction of microorganisms, but not usually bacterial spores; it may not kill all microorganisms but reduces them to a level which is neither harmful to health, nor the quality of perishable foods'.
Documentation	All the written production procedures, instructions and records, quality control procedures, and recorded test results involved in the manufacture of a product.
Extrinsic	A property derived from outside of the food product or ingredient; for example, an extrinsic hazard could arise from contamination from the environment, equipment, people or pests (see Intrinsic).
Finished Product	A product that has undergone all stages of manufacture and packaging.
Food Allergen	A food substance that, in some sensitive individuals, causes an immune response causing bodily reactions resulting in the release of histamine and other substances into the tissues from the body's mast cells in the eyes, skin, respiratory system and intestinal system. Allergic reactions may range from relatively short-lived discomfort to anaphylactic shock and death.
Food Control	The Institute of Food Science & Technology uses the term 'food control' to describe a comprehensive quality and safety system involving HACCP-linked quality assurance and quality control.

Food Hygiene	All environmental factors, practices, processes and precautions involved in protecting food from contamination by any agency, and preventing any organism present from multiplying to an extent that would expose consumers to risk or result in premature spoilage or decomposition of food.
Food Poisoning	Illness associated with consumption of food that has been contaminated, particularly with harmful microorganisms or their toxins.
Food Safety Management System	FSMS; a documented set of policies, procedures and associated documentation that combines to form a comprehensive system that effectively manages food safety.
Food Spoilage	The deterioration of food including that caused by the growth of undesirable microorganisms to high levels, which may result in fermentation, mould growth and development of undesirable odours and flavours.
Frozen Foods	Foods preserved by freezing and storing at temperatures low enough to inhibit the growth of microorganisms and to retard chemical and physical reactions to a negligible rate.
Functional	Fulfilling a specific physical, chemical or biological function.
Functional Foods	The term is one of the marketing-coined names (others are 'nutraceuticals' and 'designer foods') to categorise foods that are considered or claimed to offer specific health benefits over and above those provided by recognised nutrients, while avoiding the requirement to be licensed medicines.
Genetic Modification	GM; the process of making changes to the genes of an organism (whether an animal or plant organism or a microorganism). Genetic changes occur spontaneously in nature over a long period of time, but they may be produced intentionally either by traditional methods of selective breeding of animals and plants, or by modern methods of removal or insertion of genes. The latter method involves four basic steps:

1. the deoxyribonucleic acid (DNA) of a cell of the donor organism is broken down and the pieces separated;

2. the desired gene is selected;

3. that gene is copied many times; and

4. nth generation laboratory copies (not the donor's original genes) are then inserted into the DNA of the receiver organism.

'Within-species' genetic modification is essentially similar to traditional breeding methods (except that it is much speedier and much less haphazard). Through 'trans-species' modification, results are obtainable that could not be obtained by traditional breeding methods. In relation to food, the potential scientific benefits of genetic modification are:

- improved agricultural performance (yields) with reduced use of pesticides;
- ability to grow crops in inhospitable environments (e.g. via increased ability of plants to grow in conditions of drought, salinity and extremes of temperature);

245

- delayed ripening, permitting improvements in quality and processing advantages;
- altered sensory attributes of food (e.g. flavour, texture);
- improved nutritional attributes, for example, combating anti-nutritive and allergenic factors, and increased vitamin A content in rice; and
- improved processing characteristics leading to reduced waste and lower food costs to the consumer.

Some forms of trans-species modification may give rise to ethical and religious issues.

Genetically Modified Organism	GMO; descriptive of an organism undergoing genetic modification, or of an organism resulting from genetic modification (see above).
Good Manufacturing Practice	GMP; that combination of manufacturing and quality control procedures aimed at ensuring that products are consistently manufactured to their designated specifications.
HACCP Plan	A document that is prepared using the seven principles of HACCP, as defined in Codex Alimentarius, in order to identify realistic food safety hazards, the points at which they could arise in the manufacturing process (including materials intake and transport to the consumer) and the means for their effective control.
Hazard	A property of a system, operation, material or situation that could, if uncontrolled, lead to an adverse consequence.
Hazard Analysis	Preparation of a list of the steps in a manufacturing process (best done by preparing and verifying a process flow chart), identification of points at which hazards could arise, and then an assessment of the nature and potential seriousness of each hazard, so as to establish CCP(s) and the means for their effective control.
Hazard Analysis Critical Control Point	HACCP; a systematic preventive food safety tool designed to assist manufacturing organisations to develop an appropriate and effective food safety management system (FSMS).
HAZOP	Hazard analysis and operability study (HAZOP) is a systematic structured approach to questioning the sequential stages of a proposed operation in order to optimise the efficiency and the management of risk. Thus, the application of HAZOP to the design of a proposed food-related operation should result in a system in which as many critical control points as possible have been eliminated, making HACCP during subsequent operations much easier to carry out.
High-Care Area	HCA; an area designed to a high standard of facility specification and hygienic design where practices relating to personnel, ingredients, equipment and environment are managed to minimise microbial contamination of a RTE or RTRH product containing uncooked ingredients

High-Risk Area	HRA; an area designed to a high standard of facility specification and hygiene design where practices relating to personnel, ingredients, equipment and environment are managed to minimise microbial contamination of a RTE or RTRH product comprising only cooked ingredients.
Identity Preserved Material	A material or food product that is traceable to a known source or specific method of growing or food production, for example, kosher, halal, organic and farm assured.
Ingredients	All materials, including starting materials, processing aids, additives and compounded foods, that are included in the formulation of the product.
In-Process Control	A system of checks made and actions taken during the course of manufacture to ensure that materials at any stage comply with the specification for that stage, and that the processing and processing environment comply with the conditions stated in the master manufacturing instructions.
Intermediate Material	A partly processed material that must undergo further processing before it becomes a bulk product or a finished product.
Intrinsic	An inherent component of a food. Intrinsic food safety hazards are derived from the product, for example, bones in fish and stones in fruit. Control measures are therefore product specific and can include declarations on the packaging of the finished product.
Irradiated	Having been subjected to ionising radiation.
Low-Risk Area	LRA; an area where good manufacturing practice (GMP) standards are in place as described within this publication but the area and the practices have not been specifically designed to minimise microbial contamination, for example, raw material intake, storage areas of RTC foods and packaged product where the product is fully enclosed (see 32.19).
Manufacture	The complete cycle of production of a food or drink product from the acquisition of all materials through all stages of subsequent processing, packaging and storage to the despatch of the finished product.
Master Manufacturing Instructions	A document or documents identifying the raw materials, with their quantities, to be used in the manufacture of a product, together with a description of the manufacturing operations and procedures including identification of the plant and facilities to be used, processing conditions, in-process controls, packaging materials to be used and instructions for the removal of the finished product to storage.
Monitoring	The process of undertaking pre-scheduled sequence of observations or measurements to assess whether a product, ingredient, process, procedure, prerequisite programme or critical control point (CCP) is adequately controlled in order to consistently manufacture safe and legal food of the required quality.

Non-Conformance/ *Non-Compliance*	A failure to comply with an element of the food safety management system (FSMS), quality management system (QMS) or HACCP plan. The failure can be classed as major or minor. The term critical is sometimes used to highlight a major non-conformance that has been identified that could lead to a food safety or legality issue.
Novel (Food, *Process)*	A food or food ingredient produced from raw material that has not hitherto been used (or has been used only to a small extent) for human consumption in the area of the world in question, or that is produced by a new or extensively modified process not previously used in the production of food. Any person or company in the UK contemplating marketing a novel food or one containing a novel ingredient must make a prior submission to the Advisory Committee on Novel Foods and Processes (ACNFP) in accordance with the UK Novel Foods and Novel Food Ingredients Regulations 1997, for approval by the EU Commission. 'At what point does a novel food (e.g. mycoprotein) having come onto the market and being widely consumed, cease to be a novel food'? Novel means 'new' rather than 'unusual', and it can be argued that once a food has been through the regulatory process, it would no longer be new one year after having achieved national distribution.
Nutraceutical	See Functional foods.
Packaging	Any container or material used in the packaging of a product. This may include materials in direct contact with the product, printed packs, including labels, carrying statutory and other information, and other packaging materials including outer cartons or delivery cases. These categories are, of course, not necessarily mutually exclusive.
Prerequisite *Programme*	PRP; an element of the food safety management system (FSMS) and/or quality management system (QMS) that has been adopted by the manufacturer in order to effectively manage food safety, legality and compliance with quality specifications. The existence and effectiveness of prerequisite programmes should be considered during the development of the FSMS and HACCP plan and in the determining of critical control points (CCPs).
Preservation Index	A term deriving from the pickles and sauces industry to designate the percentage of acetic acid contained in the total volatile constituents of a product or ingredient, thus indicating probable microbial stability.
Preventive Action	An action that is adopted in order to address a weakness in a food safety management system (FSMS) and/or quality management system (QMS) that has not to date been responsible for non-conformance, but if not addressed could do so in the future.
Processed	Having been subjected to treatment designed to change one or more of the properties (physical, chemical, microbiological, sensory) of food.
Processing Aid	The UK Food Labelling Regulations 1996 define 'processing aid' as 'any substance not consumed as a food by itself, intentionally used in the processing of raw materials, foods or their ingredients, to fulfil a certain technological purpose during treatment or processing, and which may result in the unintentional but technically unavoidable presence of residues of the substance or its derivatives in the final product, provided that these residues do

not present any health risk and do not have any technological effect on the finished product'. It follows that a processing aid is an additive that facilitates processing without significantly influencing the character or properties of the finished product. Examples would be a tablet release agent used to coat the inside of tablet moulds, or a spray used to allow bread to be released from baking tins or trays. However, a substance may be a processing aid or a declarable additive, depending on its mode of use. For example, if an anti-caking agent is added to a powder ingredient to facilitate its flow properties while being conveyed to a mixer, where it is incorporated into a dough that is baked, the anti-caking agent has no technological effect in the finished bread. In this instance, the anti-caking agent is a processing aid, and hence need not be declared. If however, that powder ingredient is directly packed into containers for sale as such, or is incorporated in a dry mix product, the anti-caking agent continues to have an anti-caking effect in that product. Hence it is not acting solely as a processing aid and must be declared as an additive.

Quality Assurance	See Chapter 2.
Quality Control	See Chapter 2.
Quality Management System	QMS; the organisational framework of policies, procedures, associated documentation and resources needed to implement the strategy required to consistently deliver product that is within predetermined specifications.
Quarantine	The status of any materials or product set aside while awaiting a decision on its suitability for its intended use or sale.
Radiation Dose	Doses of radiation are defined in terms of the energy absorbed by the substance irradiated. The unit for radiation dose is the gray (Gy), which is defined as the dose corresponding to the absorption of 1 joule per kilogram of the matter through which the radiation passes [1 kilogray (kGy) = 1000 Gy].
Raw Material	Any material, ingredient, starting material, semi-prepared or intermediate material, packaging material and so on used by the manufacturer for the production of a product.
Ready to Cook	RTC; food designed to be given a heat process by the consumer that will deliver a 6-log kill with respect to vegetative pathogens (a minimum process equivalent to 70°C for 2 minutes) throughout all components.
Ready to Eat	RTE; food intended by the producer or the manufacturer for direct human consumption without the need for cooking or other processing effective to reduce to an acceptable level or eliminate microorganisms of concern (i.e. cold eating).
Ready to Reheat	RTRH; food manufactured in a high-care area (HCA) or high-risk area (HRA) that has been designed to be reheated by the final consumer.
Re-Validation	The process by which an element of the food safety management system (FSMS), quality management system (QMS) or HACCP plan is reassessed to ensure is continuing ability to deliver food safety, quality or legal compliance objectives.

Reworking	The process of taking food that does not meet specification and either reprocessing, resorting or otherwise handling in order to address the non-conformance so that it will then meet specification (see Chapter 21).
Risk	The probability that a particular adverse consequence results from a hazard within a stated time under stated conditions.
Risk Assessment	The management process of identification, evaluation and estimation of the levels of risk associated with a food safety hazard, situation or process or procedural failure. Further categorisation of the risk identified is by determining the likelihood and the severity of occurrence and determining the acceptable level of risk to the consumer. This can include the acceptable level of a component in a food product, for example, a pesticide residue, or absence versus presence of a foreign body.
Root Cause Analysis	The structured management approach that identifies the factors that resulted in non-conformance in order to determine the most appropriate corrective or preventive action. The factors that could be considered include the actual nature of the non-conformance, the magnitude (major or minor) and the consequences of the problem in order to identify the actions, conditions or behaviours that need to be changed to prevent reoccurrence and/or other similar problems from occurring.
Specification	A document giving a description of material, machinery, equipment, process or product in terms of its required properties or performance. Where quantitative requirements are stated, they are either in terms of limits or in terms of standards with permitted tolerances.
Starting Material	See Raw material.
Validation	The process of obtaining of evidence that the elements of the food safety management system (FSMS), quality management system (QMS) or HACCP plan are effective at delivering food safety, quality or legal compliance objectives.
Verification	The process of developing procedures, assessments and other evaluations including auditing, that is, in addition to monitoring to determine compliance with the food safety management system (FSMS), quality management system (QMS) and HACCP plans.

APPENDIX II

ABBREVIATIONS USED IN THE GUIDE

Note: In some instances, an abbreviation of the former name of an organisation is included, where a document referred to was issued by, and bears the former name of, that organisation.

ACNFP	Advisory Committee on Novel Foods and Processes (UK)
ALOP	Acceptable level of protection
APC	Aerobic plate count
ATP	Adenosine triphosphate
AVA	Automatic Vending Association [formally the Automated Vending Association of Britain (AVAB)]
BPCA	British Pest Control Association
BRC	British Retail Consortium
BS	British Standard
BSI	British Standards Institution
CAC/RCP	Codex Alimentarius Commission/Recommended Code of Practice
CAC/RS	Codex Alimentarius Commission/Recommended Standard
CCP	Critical control point
CCTV	Closed-circuit television
CEN	Comité Européene de Normalisation
CFA	Chilled Food Association
CIP	Cleaning in place
COO	Country of origin
COSHH	Control of Substances Hazardous to Health
CPD	Continuous professional development

Food & Drink – Good Manufacturing Practice: A Guide to its Responsible Management, Sixth Edition.
The Institute of Food Science & Technology Trust Fund.
© 2013 John Wiley & Sons, Ltd. Published 2013 by John Wiley & Sons, Ltd.

CQP	Critical quality point
CSF	Critical success factor
DEFRA	Department of Environment, Food and Rural Affairs (UK)
EC	European Community
EDP	Electronic data processing
EFK	Electric fly killer (or insectocutor)
EFSA	European Food Safety Authority
EN	Denotes a regional standard intended to be used in the European Union
ETI	Ethical Trading Initiative
EU	European Union [formerly European Community (EC) and originally Economic Community (EEC)]
EVA	European Vending Association
FAO/WHO	Food and Agriculture Organisation/World Health Organisation (United Nations)
FAPAS	Food Analysis Performance Assessment Scheme
FEPAS	Food Examination Performance Assessment Scheme
FDA	Food and Drug Administration (USA)
FIFO	First-in-first-out
FMA	Food Machinery Association (UK)
FMEA	Failure mode and effects analysis
FMF	Food Manufacturers' Federation (UK)
FRC	Free residual chlorine
FSA	Food Standards Agency
FSMS	Food safety management system
GFSI	Global Food Safety Initiative
GLA	Gangmaster Licensing Authority
GLP	Good laboratory practice

GMO	Genetically modified organism
GMP	Good manufacturing practice
HACCP	Hazard analysis critical control point
HAZOP	Hazard analysis and operability study
HCA	High-care area
HMSO	Her Majesty's Stationery Office (now The Stationery Office)
HPA	Health Protection Agency
HRA	High-risk area
HSE	Health and Safety Executive
ICCT	International Cold Chain Technology
ICMSF	International Committee for Microbiological Specifications for Foods
IEC	International Electrotechnical Commission
IFST	Institute of Food Science & Technology (UK)
IPPC	Integrated Pollution Prevention and Control
ISO	International Organization for Standardization
ISO/TS	International Organization for Standardization/Technical Specification
JECFA	Joint FAO/WHO Expert Committee on Food Additives
LACORS	Local Authorities Coordinators of Regulatory Services (UK) [formerly LACOTS (Local Authorities Coordinating Body on Trading Standards)]
LRA	Low-risk area
MAFF	Ministry of Agriculture, Fisheries and Food (UK)
MAP	Modified atmosphere packaging
MITE	Measuring inspection and testing equipment
MRA	Microbiological risk assessment
NAMAS	National Measurement Accreditation Service (see UKAS)
NPTA	National Pest Technicians Association

OECD	Organisation for European Co-operation and Development
PAS	Publicly Available Specification
PPE	Personal protective equipment
PRP	Prerequisite programme
RCP	Recommended Code of Practice
RTC	Ready to cook
RTE	Ready to eat
RTP	Returnable transit packaging
RTRH	Ready to reheat
Q	Qualitative
QCP	Quality control point
QMS	Quality management system
QRA	Quantitative risk assessment
SA	Social accountability
SEDEX	Supplier Ethical Data Exchange
SI	Statutory Instrument (UK)
SMETA	Sedex Members Ethical Trade Audit
SPS	Sanitary and Phytosanitary
SQ	Semi-quantitative
UKAS	UK Accreditation Service (formerly NAMAS, q.v.)
USA	United States of America
UV	Ultraviolet
VR	Verification risk
WHO	World Health Organisation
WTO	World Trade Organisation

APPENDIX III

LEGISLATION AND GUIDANCE

There are a vast number of guides and industry codes of practice issued by a variety of bodies, some of which are referenced in the previous chapters. Previous editions of the Guide have included a comprehensive listing of such guides and codes. These will not be repeated in this appendix; instead, a list of websites is identified where further information can be accessed.

The 'Food Law Guide' is issued by the Food Standards Agency (FSA), Room 245, Aviation House, 125 Kingsway, London WC2B 6NH and is an invaluable listing of UK Food Laws and Regulations, with brief descriptive outlines and indication of European Union (EU) Directives to which each gives effect in the UK. The Guide is updated quarterly and also available on the FSA website. To take account of legislative provisions, reference should be made to the actual texts of the relevant laws and regulations, purchasable from The Stationery Office or any of its bookshops (http://www.tso.co.uk).

Websites

Automatic Vending Association
http://www.ava-vending.org

British Frozen Food Federation
http://www.bfff.co.uk

British Pest Control Association (BPCA)
http://www.bpca.org.uk

British Standards Institution
http://www.bsigroup.co.uk

Campden BRI
http://www.campden.co.uk

Chilled Food Association (CFA)
http://www.chilledfood.org

Codex Alimentarius Standards
http://www.codexalimentarius.net/web/standard_list.do?lang=en

Department of Food, Environment and Rural Affairs (UK)
http://www.defra.gov.uk

Food & Drink – Good Manufacturing Practice: A Guide to its Responsible Management, Sixth Edition.
The Institute of Food Science & Technology Trust Fund.
© 2013 John Wiley & Sons, Ltd. Published 2013 by John Wiley & Sons, Ltd.

European Union: European Food Safety Authority
http://www.efsa.europa.eu/

European Vending Association
http://www.eva.be

Food and Agriculture Organisation (FAO)
http://www.fao.org

Food Standards Agency
http://www.food.gov.uk

IFST Information Statements
http://www.ifst.org/science_technology_resources/for_food_professionals/
information_statements/

Institute of Food Science & Technology (IFST)
http://www.ifst.org

International Commission on Microbiological Specifications for Foods
(ICMSF)
http://www.icmsf.org/index.html

National Pest Technicians Association
http://www.npta.org.uk

The Royal Society
http://www.royalsoc.ac.uk

The Stationery Office
http://www.tso.co.uk

World Health Organisation (WHO)
http://www.who.int

APPENDIX IV

ADDITIONAL REFERENCES

This appendix provides some references to topics relevant to good manufacturing practice (GMP) and its management. The sheer volume of the publications available makes it impractical for the list to be exhaustive either in scope or within subject.

BRC Global Standard for Food Safety: Issue 6 The British Retail Consortium. TSO (The Stationery Office) July, 2011. Other guidance available at the TSO website.

BS EN 1276:1997. Chemical disinfectants and antiseptics. Quantitative suspension test for the evaluation of bactericidal activity of chemical disinfectants and antiseptics used in food, industrial, domestic, and institutional areas. Test method and requirements (phase 2, step 1). December 1997.

BS EN 13697:2001. Chemical disinfectants and antiseptics. Quantitative non-porous surface test for the evaluation of bactericidal and/or fungicidal activity of chemical disinfectants used in food, industrial, domestic and institutional areas. Test method. September 2001.

BS EN 1499-1997. Chemical disinfectants and antiseptics – Hygienic handwash – Test method and requirements (phase 2/step 2). Replaced Standard: 94/504554 DC-1994. October 1997.

Campden BRI produce a large range of books, guidelines and research reports for every area of food manufacturing. Available at http://campden. co.uk/publications/pubs.php

The Campden BRI Guideline G48: Guidelines for preventing hair contamination of food, 2006. Available at http://www.campden.co.uk/publications/ pubDetails.php?pubsID=98

The Campden Guideline G62: Hand hygiene: Guidelines for best practice, 2009. Available at http://www.campden.co.uk/publications/pubDetails. php?pubsID=4480

The Chilled Foods Association produce a range of guidelines. Available at http://www.chilledfood.org/resources/publications

Food & Drink – Good Manufacturing Practice: A Guide to its Responsible Management, Sixth Edition.
The Institute of Food Science & Technology Trust Fund.
© 2013 John Wiley & Sons, Ltd. Published 2013 by John Wiley & Sons, Ltd.

These include:

- Best Practice Guidelines for the Production of Chilled Food (4th Edition). The Stationery Office, London, 2006.
- BRC/CFA Guidance on the Practical Implementation of the EC Regulation on Microbiological Criteria for Foodstuffs. http://www.chilledfood.org/content/guidance.asp
- Microbiological Guidance for Produce Suppliers to Chilled Food Manufacturers. Micro Guidance for Growers (2nd Edition), 2007.

Chilled Foods: A Comprehensive Guide (3rd Edition). Ed. Brown, M., Woodhead Publishing, 2008.

Code of Practical Guidance for Packers and Importers. Weights and Measures Act 1979 Issue No. 1. Department of Trade and Industry. DTI. TSO (The Stationery Office). ISBN 9780115129223.

Crisis Management in the Food and Drinks Industry: A Practical Approach (2nd Edition). Ed. Doeg, C., Springer, 2005.

Detecting Foreign Bodies in Food. Ed. Edwards, M., Woodhead Publishing, 2004.

Fermented Beverage Production (2nd Edition). Ed. Lea, A.G.H. and Piggoitt, J.R., Springer, 2003.

Food Safety for the 21st Century: Managing HACCP and Food Safety Throughout the Global Supply Chain. Ed. Wallace, C., Sperber, W. and Mortimore, S.E., Wiley-Blackwell, 2010.

Food Standards Agency. *E. coli* O157—Control of cross-contamination: Guidance for food business operators and enforcement authorities, 2011. Available at http://www.food.gov.uk/business-industry/guidancenotes/hygguid/ecoliguide#.UFsKU42PXWE

Food Standards Agency. Food handlers: Fitness to work regulatory guidance and best practice advice for food business operators, 2009. Available at http://www.food.gov.uk/multimedia/pdfs/publication/fitnesstoworkguide09v3.pdf

Food Standards Agency. Food Law Code of Practice (England), 2012. Available at http://www.food.gov.uk/multimedia/pdfs/codeofpracticeeng.pdf

Food Standards Agency. Food Law Practice Guidance (England), 2012. Available at http://www.food.gov.uk/multimedia/pdfs/practiceguidanceeng.pdf

Foreign Matter Prevention and Detection: A Practical Approach. Ed. Haycock, P.J. and Wallin, P.J., Blackie Academic & Professional, 1998.

HACCP: A Practical Approach (3rd Edition). Ed. Mortimore, S.E. and Wallace, C., Springer, 2000.

Handbook of Hygiene Control in the Food Industry. Ed. Lelieveld, H.L.M, Mostert, M.A and Holah, J., Woodhead Publishing, 2005.

Handbook of Organic Food Processing and Production. Ed. Wright, S., Blackie Academic & Professional, 1994.

The Principles of Design for Hygienic Food Processing Machinery. Technical Memorandum 289. CCFRA, 1982.

Principles of Food Sanitation (5th Edition). Ed. Marriott, N.G. and Gravani, R.B., Springer, 2006.

Principles and Practices for the Safe Processing of Foods. Ed. Shapton, D., CRC Press, 1998.

Quality Assurance for the Food Industry: A Practical Approach. Ed. Vasconcellos, J.A., CRC Press, 2003.

Shelf Life. Ed. Man, D., Blackwell, 2002.

Statistical Quality Control for the Food Industry (3rd Edition). Ed. Hubbard, M.R., Springer, 2003.

Total Quality Assurance for the Food Industries (3rd Edition). Ed. Gould, W.A. and Gould, R., Woodhead Publishing, 2001.

Total Quality Management: The Route to Improving Performance (2nd Edition). Ed. Oakland, J.S., Butterworth-Heinemann, 1994.

APPENDIX V

LIST OF ORGANISATIONS AND INDIVIDUALS FROM WHOM HELP, INFORMATION OR COMMENT HAS BEEN RECEIVED

GMP Working Group   Dr Louise Manning (Editor)
Members   IFST Scientific Committee members (including Prof. J.R. Blanchfield)
    Miss K.E. Goodburn
    Argyll and Bute Council

This 6th edition has, of course, built on and updated the valuable content of previous editions, and substantially reflects the expertise and efforts of all those who contributed in various ways to one or more of the five editions. It is appropriate that they are all recorded here.

Members of Working   Prof. K.G. Anderson
Groups for Previous   Mr R.J.W. Anderson
Editions   Mr P.L. Bidder
    Mr J.R. Blanchfield
    Prof. G. Campbell-Platt
    Prof. C. Dennis
    Mr P.O. Dennis
    Mr E. Druce
    Dr E. Green
    Miss K.A. Herrington
    Mr M.C.K. Kane
    Dr E. Moss
    Mr P. Berry Ottaway
    Dr D. Simpson
    Mr A.J. Skrimshire
    Mr R.L. Stephens
    Dr T.G. Toomey
    Mr A. Turner
    Mr W.E. Whitman
    Mr S. Wood

Thank you too for all the contributions, help, advice and comments received from individuals and organisations in the previous five versions.

Food & Drink – Good Manufacturing Practice: A Guide to its Responsible Management, Sixth Edition.
The Institute of Food Science & Technology Trust Fund.
© 2013 John Wiley & Sons, Ltd. Published 2013 by John Wiley & Sons, Ltd.